French Inventions
of the
Eighteenth Century

French

OF THE

By

University

Inventions

EIGHTEENTH

CENTURY

Shelby T. McCloy

of Kentucky Press

COPYRIGHT, 1952, BY THE
UNIVERSITY OF KENTUCKY PRESS

PRINTED AT THE
UNIVERSITY OF KENTUCKY BY
THE KERNEL PRESS

Library of Congress Catalog Card Number:
52-5903

To my parents

Joseph Dixon McCloy
and
Sarah Tommie Tool McCloy

Preface

THE PURPOSE OF THIS BOOK IS TO SET FORTH AN ACCOUNT OF French inventive activity during the 1700's and to give the reader a clearer idea of what the French were doing prior to and during the period when the British entered their important Industrial Revolution. Save for the accounts of a few French inventors, like Cugnot, Chappe, Vaucanson, Pinel, and Coulomb, the story of French invention of the eighteenth century is largely unknown. Not only is there no book in English covering the subject, there is none even in French treating all its aspects. Two French works deal with parts of it in excellent fashion. One is the four-volume *Merveilles de la science*, by Louis Figuier, prolific nineteenth-century historian of science, published in the late 1860's. It rests upon minute research and is written entertainingly. Hundreds of illustrations add to its attractiveness. The other work is the doctoral treatise of Charles Ballot, *L'introduction du machinisme dans l'industrie française*, published posthumously in 1923 by his friend and editor Claude Gével. Ballot, a young French scholar in his forties, was killed at Verdun in 1916. War called him away from his studies and prevented the completion of his manuscript. The writing was finished by Gével from notes that Ballot had left. The book is of the highest scholarship and is the only work in its field describing the textile developments in France in the late 1700's and early 1800's. To both Ballot and Figuier I am heavily indebted for information.

It is not my aim to set forth an exhaustive treatise. That would require years of research by one who would have to be expert in several of the sciences. My more modest design is to survey the whole field and to give a readable account of all the inventions that seem to me worthy of description. In doing this I have gleaned extensively from secondary works, but I have also made much use of primary material, such as *Histoire de l'Académie royale des Sciences, Machines et inventions approuvées par l'Académie royale des Sciences,* and *Réimpression de l'ancien Moniteur.* Much information has been obtained from special articles in learned journals.

In doing my research I have worked in the University of Kentucky Library, the Duke University Library, the Library of Congress, and the Transylvania College Library. I have also drawn upon the resources of the libraries of Ohio State University, Western Reserve University, University of Illinois, University of Minnesota, Harvard University, Leland Stanford University, Yale University, University of North Carolina, and the Surgeon General of the United States Army. To the officials of these libraries I express appreciation for their kindness.

To M. Loiseau, custodian of the National Museum of the Arts and Trades, Paris, I am grateful for providing me with many photographs of objects in his institution and for permitting the reproduction of those included herein. To the Library of Congress, the Ohio State University Library, and the University of Kentucky Library I am indebted for the other illustrations in the book.

To the Social Science Research Council I am indebted for a grant-in-aid in my study of "The Humanitarian Movement in Eighteenth-Century France."

Among others to whom I am obligated, I thank Dr. Gladys Smithwick and Professor Lyle Dawson for reading the chapters on medicine and chemistry respectively, and Professor Dewey Steele for reading the entire work, and for their criticisms; my wife, Minnie Lee McCloy, Miss Elsie Hurt, and Miss Jean Coleman for typing the manuscript; and Miss Allene Ramage for checking some references in the Duke University Library.

SHELBY T. MCCLOY

Lexington, Ky.
August 27, 1951

Contents

Preface		vii
Introduction		1
I.	The Balloon	11
II.	Steam Transportation	28
III.	The Telegraph	41
IV.	Lighting	50
V.	Papermaking	62
VI.	Chemical Inventions	71
VII.	Textiles	90
VIII.	Automata	103
IX.	Other Mechanical Devices	111
X.	Military Inventions	136
XI.	Medicine and Surgery	148
XII.	Patents and Encouragement	170
Conclusion		186
Bibliography		199
Index		205

Illustrations

	Facing page
THE STEAM TRUCK OF NICOLAS CUGNOT	6
THE FIRST PASSENGER-CARRYING HOT AIR BALLOON	14
THE FIRST PARACHUTE JUMP	20
THE FIRST SUCCESSFUL STEAMBOAT	34
THE INTERIOR OF A SEMAPHORE TELEGRAPH TOWER	46
THE CARCEL LAMP	56
THE "FLOATING BATTERIES" ATTACKING GIBRALTAR	74
THE LOOM OF JACQUES VAUCANSON	96
THE PLANISPHERE OF ABBE OUTHIER	108
THE TORSION BALANCE OF CHARLES COULOMB	118
AN INCUBATOR FOR HATCHING CHICKENS	130
THE BREECH-LOADING CANNON OF DE LA CHAUMETTE	142
A DREDGE FOR CLEANING HARBORS	156
THREE AUTOMATA OF JACQUES VAUCANSON	174
THE EXTERIOR OF A SEMAPHORE TELEGRAPH TOWER	188

Introduction

IN THE RAPID SOCIAL ADVANCE OF THE LAST TWO AND A HALF centuries, science and invention have played a gigantic role. There is no one living today, at least in a civilized country, who is not very much their debtor. At every turn we enjoy conveniences and advantages that even the most highborn and favored of the eighteenth century did not dream of. Our homes and offices have been transformed by artificial lighting and heating, hydraulic accommodations, close-fitting windows, and numerous other devices not known in that day. Communication and entertainment have been revolutionized through radio, television, the telephone, and the telegraph. Transportation has been so altered by the automobile, the locomotive, the Pullman car, the steamship, and the airplane, that the problems of distance largely have been eliminated and the traveler may journey almost as speedily as his fancy may wish and pocketbook permit. Agriculture is carried on today in an entirely different manner, with motorized machinery, chemical fertilizers, and insecticides; banking and financial institutions have been able to multiply their services a hundredfold through elaborate calculating machines; packing houses and food stores, through canning, refrigeration, and rapidly transported foodstuffs, have completely altered the diet of the public and placed it, save possibly for periods of war and civil disturbance, free from the former specter of famine. This list of the transformations in human activities could be greatly extended. If it were possible for one who lived in the eighteenth century to return to the world today, he would be baffled at every turn by what he would see and hear.

The historian of course is interested in these changes and what has caused them. Distinctly it has been the work of invention and science which has transformed the world for man even more than the fabled lamp of Aladdin. It is perhaps best to speak of the agents as twofold, for the majority of inventors have not been scientists and similarly most scientists have not been inventors. Both groups, however, employ the same fundamental method of experimentation, and the work of both is founded on the scientific knowledge and concepts of their day. As Professor John U. Nef has stated in an excellent article in the *Journal of Economic History* for 1943, the inventive achievements of the last few centuries have been consequent to and predicated on the remarkable developments in pure science. Much scientific development of course antedated the eighteenth century. Great contributions had been made by Copernicus, Galileo, Vesalius, Gilbert, Kepler, Newton, Huygens, Von Guericke, Leibniz, Papin, Pascal, Harvey, and others.

During the 1600's there had been an efflorescence of scientific academies in various cities of Italy and Germany and in Paris and London. Those of Paris and London were particularly influential. Formed in each instance in the 1660's, with government recognition and support, they both at once became distinguished and election to their membership was regarded as a mark of the highest recognition, commonly reserved for those who had made some notable contribution in science. Huxley has been quoted by the late Preserved Smith (*History of Modern Culture*, Vol. I) to the effect that if all other printed records of our modern civilization were destroyed it would be possible largely to reconstruct our knowledge of its essential features from the London academy's periodical, the *Philosophical Transactions of the Royal Society of London*. Perhaps the same might be said of the *Mémoires* of the French Royal Academy of Sciences, to which reference so frequently is made in this study. The Academy, chartered in 1666 by Louis XIV and Colbert, consisted at the outset of twenty-one members, pensioned by the government that they might devote their entire time to science. Their meeting place and equipment at the Royal Library also were furnished by the government, and the expense of publication of results was defrayed. In addition to the full members of the Academy (around forty in the 1700's), there were associates and assistants, like members of an American university faculty. Indeed, a member might undertake the instruction of a prominent young scientist, such as Réaumur, in fashion not dissimilar to the training in the guilds.

In this Academy were gathered the most brilliant men in French science, and among them some were eminent inventors.

Science in the 1700's no longer had to fight the battle for recognition. All classes of society acclaimed it. Almost every city of consequence had its academy, which, while not devoted exclusively to science in most cases, had the promotion of science among its objectives. Frequently these academies sponsored publications and offered prizes for the best papers to be submitted on some designated topic. Occasionally the meetings of these academies were open to the public. The public also was invited to other scientific lectures, such as those given in the 1780's under the auspices of the Société Apollonienne. In France some fifteen agricultural societies existed on the eve of the Revolution, and in every leading country of Western Europe great attention was given to the claims of scientific agriculture and animal breeding. Arthur Young, whose *Travels . . . During the Years 1787, 1788, 1789* are so well known, was in France to obtain information on French practices for his journal of scientific agriculture. For twenty years or more Young traveled about in the British Isles, France, Spain, Switzerland, and Italy with this end in view, partly meeting his expenses by writing travel books on his experiences. The Agricultural Revolution was well under way, above all in England where it originated in the late 1600's.

Eminent scientists frequently were courted and engaged by the heads of states. The great Swiss mathematician Euler was constantly in the employ of Russia or Prussia, and one or two of the famous Bernoulli family were also employed by Russia, serving as scientific consultants or engineers. Catherine the Great courted D'Alembert and tried without success to obtain his services as tutor for her son Paul, and Frederick the Great had the mathematician Maupertuis to head his academy in Berlin. Eminent scientists were everywhere duly recognized and acclaimed, and science in the 1700's made great advances.

Invention, too, during the century was increasingly popular, although there were periods of unpopularity in all countries when machines were smashed by rioters as agencies of unemployment. It is surprising how many of the conveniences of today which have come through invention were being sought in the eighteenth century by would-be inventors. There were the threshing machine, steamboat, truck, electric telegraph, hot-water heating system for homes, artificial lighting by gas, beet sugar, and the preservation of food by canning, to name only a few. Inventions were being attempted in almost every field of endeavor. Britain and France were the countries most active, but much invention was found also in Germany, Switzerland, Sweden, and elsewhere.

British inventors were in the forefront. Their inventions provide one of the most glorious chapters of British history, for in no small degree they laid the foundations of the remarkable advances of Britain in the nineteenth and twentieth centuries, not only in industry and commerce but also in every other sphere of activity based upon them. In textiles, farm machinery, animal breeding and agriculture, horology, glass and pottery, scientific instruments, nautical devices, metallurgy, gas lighting, medicine and surgery, and even other fields, British inventors made brilliant discoveries. If Britain had made a contribution in no other respect than in her discoveries of treatment of scurvy and smallpox, subsequent ages would have been her debtor; but she made contributions perhaps as notable in many other spheres.

France, if she was behind Britain—and I am reductant to think that she was—was certainly not far behind. In almost every sphere of human activity her gifted intellects were making contributions of great value.

Germany was a slow third in creative activity. She was not yet united politically but divided into more than three hundred states, and while she had thirty-six or more universities and many learned academies, she did not strike her stride of creative brilliance until the nineteenth century. Nevertheless, in the fields of chemistry, metallurgy, scientific instruments, printing of dry goods and paper, lithography, electricity, surgery, and other matters, German inventors made contributions.

The Swiss were a small people, but they produced men of eminence in various fields in the 1700's. In invention they were remarkably active, above all in horology, glassware, textiles, scientific instruments, and automata. The elder Savery, the two Drozes, Guinand, Deboule, and Berthoud were a few of the notable Swiss inventors.

The remarkable achievements of Scheele in chemistry and of Linnaeus in botany, while lying more in the realm of pure science than of invention, have nevertheless some claim to the latter and made the name of Sweden famous in the 1700's. Scheele created several dozen new chemical acids and salts and in other respects advanced the field of chemistry in a dazzling manner. Linnaeus created a new mode of classification for plants and a binomial terminology of botanical subjects which was considered a marked improvement over previous ones, and is commonly regarded as the Father of Modern Botany.

The Italians made some interesting inventions in the fields of electricity and medicine, as did also the Dutch. The latter also continued their seventeenth-century role of distinction in the making of scientific instruments.

As for American inventions, we can cite those by Franklin in electricity, his lamp and stove, the steamboat of John Fitch and James Rumsey, the cotton gin of Eli Whitney, a carding machine of Oliver Evans, and a grain drill of Eliakim Spooner. Perhaps there were others of merit, but American invention has lain mostly in the nineteenth and twentieth centuries.

Hitherto, British inventions in cotton textiles and Watt's improvement of the steam engine have attracted paramount attention, as they formed the vanguard of that great movement known as the Industrial Revolution, one of the most significant forces in modern history. And yet these several inventions which rapidly transformed the British cotton industry after 1770 or 1780 were but a trivial portion of the multitude that the century brought forth.

For this great outburst of inventive activity several factors were responsible, the foremost of them being the great advance of science. The pure scientist was in the vanguard; the inventor was not far behind. Outstanding among other factors operative was that of education. Some of the inventors were men of little schooling, but most of them were well educated. All were well trained technically in their field of endeavor. Many had received their training as apprentices. This included not only the technicians proper but also the surgeons and some of the scientists. During the second half of the 1700's there was a remarkable development of technical schools, above all in France and to a lesser degree in the other countries of Western Europe. Schools for engineering, mining, veterinary science, military and naval methods, and instruction of deaf-mutes and the blind flourished rather generally, usually with government support. They gave expression to the strong undercurrent of dissatisfaction voiced by the *philosophes* and others that education should deal with "practical" studies and less with Latin and Greek. In its desire to emphasize these practical phases of education, the French government not only subsidized the instruction but gave scholarships on a wide and liberal scale to the students. Most of the inventors were men of education, and education as a whole throughout Western Europe was increasing among the masses, though far from being universal. Of education as of science it might be said that as it has increased in the last few centuries, invention has followed in its wake.

Government encouragement in one form or another, at least in France, was a very important factor in promoting invention. In Britain much the same end was achieved by private means. In both countries inventors were granted patents (the term in France was "brevet") and they were allowed a period, commonly fifteen years, for the economic exploitation of their inventions. In France, however, there were handicaps during the first half of the century in that the government granted many commercial and industrial monopolies to favorites and held to rigid enforcement of the narrow guild regulations. From the guilds happily the British had emancipated themselves, and this freedom was a boon to British invention no less than to British economic enterprise. During the second half of the century, as Professor Nef has shown, a wiser policy prevailed in France, and not only did the government remove the monopolies by favorites and grow lax in the enforcement of guild regulations, but also it went in heavily for industrial and inventive subsidies.

War, too, was a factor of importance. Many were the inventions in firearms, cannon, and explosives, as will be discussed in Chapter X. Nor was this all. War stimulated the development of ballooning, which came to be applied even in the 1790's to warfare, and it appears that a war-created shortage of paper in

THE STEAM TRUCK OF NICOLAS CUGNOT

The large covered container in front holds both the firebox and the steam. This particular machine was Cugnot's second, made in 1771.

(Courtesy of the Conservatoire National des Arts et Métiers, Paris, where it rests)

France suggested to Robert the need of manufacturing paper by machine. The steam truck, invented by Cugnot, was definitely created for a military purpose, although fulfillment of that end was not attained until 1914. Pontoon bridges and other devices arising out of the needs of war received the attention of inventors. War cramped certain types of activity but encouraged others, even as it has always absorbed a great portion of man's attention and endeavor.

Last but perhaps not least among the explanatory factors was the keen commercial and industrial rivalry of the times, which then as today had its local and its international aspects. Trade with the Orient in particular was great and not only played its part in international war but also in influencing taste and costume and manufacture in Western Europe. The Delft and Meissen wares, the wares of Sèvres, Chelsea, and Staffordshire, the painted cotton goods of the late 1700's, and lacquers were creations arising from this competition. The rivalry between Britain and France had a gigantic part to play in French invention and industrial development in the second half of the century, when the government set out purposely and without stint to copy British methods and overhaul Britain's industrial leadership. Within each country there was then, as now, the perpetual striving of industrialists and others to gain advantage wherever possible over rivals by improved techniques. This condition certainly was operative on a large scale in Britain, as has been pointed out by every writer on the subject, and no doubt it was also important in France.

The eventual culmination of these forces, at least in manufactures, was what we term the Industrial Revolution, a movement that began in Britain about 1770 or 1780 and developed sufficiently to be of enormous aid to her in the contest with Napoleon in the early 1800's. France was on the point of experiencing it when the great political and social Revolution of 1789 broke out, diverting attention largely from industrial development, with the result that the Industrial Revolution there was delayed until the 1820's.

It is all the more remarkable that France did not experience the revolution in industry earlier, along with Britain, when we consider the role that she played in European affairs. France of the 1700's was strong politically, economically, culturally. In a real way she was the center of learning and enlightenment. Visitors from all over the world came there to see the latest and the best in manners, dress, the fine arts, science. Students from Britain, Germany, Scandinavia, and elsewhere came with their tutors on the Grand Tour, for no one's education was quite complete without a stay in Paris. The French language was the language of cultured society throughout Europe, and Frederick the Great, Catherine the Great, and other rulers of the period used it commonly. Washington, Jefferson, Franklin, and other prominent Americans used French as a second language and partly filled their libraries with the works of French thinkers. Large numbers of foreigners settled in France as their adopted home. There were heavy immigrations of Germans, Swiss, Dutch, Belgians, Irish, English, Scots, and Belgians, and smaller numbers of other nationalities. Many of these were hired outright by the government for military service or industrial development; others, as agricultural workers and colonists, it encouraged directly or indirectly. Perhaps a fifth of the French army was composed of foreigners. In it were numerous regiments of Germans and Swiss, and several regiments of other nationalities. There was one of Turks and Negroes. These foreign regiments, it will be recalled, played a notorious role in the Revolution, as they were more loyal to the king than were most of the native French troops. A number of prominent industrialists and inventors of eighteenth-century France were of foreign birth, but on the other hand France through her Protestant refugees contributed a considerable number to other countries.

France's government was an absolutism, but in many respects a benign one. Throughout the century it displayed an interest in industry and invention, not always wisely, but during the second half it took a more vigorous and effective part. This awakened interest the Revolutionary and Napoleonic governments continued, though not always in the same respects or with the same success. Their problems came to be dictated largely by military necessity, and the demands of a country at war are in many ways different from those of a country at peace. Some industries, like chemicals and munitions, developed greatly, while others, like textiles, felt a slackening.

French glory did not end with the eighteenth century. French scientists and inventors have made hundreds of contributions since 1800, as may be seen by reading Léon Guillet's *La France, pays de grandes découvertes* (Paris, 1947), and French industrialists have taken an active part in world production of goods, more especially in chemicals, cosmetics, soaps, metallurgy, textiles, machines, and munitions. Of French inventors, only a few are known to the world at large, as Jacquard for his loom, Appert for his canning, Niepce and Daguerre for photography, Pasteur for his antibodies, Braille for his printing, and the Curies for radium. These discoveries and inventions alone were of colossal import when one considers their effect upon the life of today. Of course most of the French inventions and discoveries since 1800 have had less significance, and yet they have been important. To name a few of the lesser known inventions, the typewriter, by Xavier Progin, has become indispensable, and neon lighting, by Georges and André Claude, is rapidly coming to be so, too. The first submarine periscope and the first Diesel motor for submarines are credited to Jean Rey (1861-1935). Photolithography (1851) was created by Lemercier, and photoengraving (1876) by Firman and Charles Gillot. The gyroscope was invented by Léon Foucault (1852), and the helicopter (1907) by Louis and Jacques Bréguet and Professor Richet. Auguste Rateau is credited with a number of motor inventions for automobiles and aeroplanes. In fact, a number of French inventors made contributions for the propulsion not only of automobiles and aeroplanes but also of submarines and surface vessels. To go further with citations would perhaps only be wearisome. The interested reader may consult Guillet.

Unfortunately Guillet is in French, and while French is perhaps the most widely known of the foreign languages in America, this constitutes a great barrier. Even those who can read a foreign language rarely will do so. It is this which, in my opinion, makes the history of France much less known in America than the history of England. For the history of most countries has quite naturally been written by their nationals. Only the chefs-d'oeuvre are translated, and special studies of French history by British and American scholars are surprisingly few. Moreover, the French, like other nations, have not recorded adequately their own history of science and invention. One is amazed at the neglect.

It would probably be erroneous to predicate a nation's power or wealth wholly upon its inventions. Even as one inventor prospers and another dies in poverty, so to some extent countries also fare. I would not attribute the outstanding role of Soviet Russia to scientific or inventive genius. Other factors are involved. Likewise in trying to explain the less resplendent role of France in the twentieth century, I should say that population has been a paramount factor. In the eighteenth century she had a larger population than most European states save Russia. Germany was disorganized, and Russia still slumbered in ignorance and lethargy. France today is badly outnumbered by both. Moreover, it must not be forgotten that since 1870 she has suffered three Germanic invasions. They have told heavily on her, each being progressively severe. The fear of their repetition has impelled the French for seven decades to maintain military and naval preparedness to a degree that perhaps has precluded development in other national aspects; and as long as danger lies in Germany or in Russia there can be no change in the future.

CHAPTER I

The Balloon

A FAMOUS DAY IN THE ANNALS OF THE SMALL TOWN OF Annonay, near Lyons in southeastern France, was June 5, 1783, when a group of its citizens watched the first public balloon ascension.[1] The heroes of the occasion were the inventors Joseph (1740-1810) and Etienne (1745-1799) Montgolfier, sons of a well-to-do paper manufacturer.

According to report, the Montgolfier brothers first became interested in the idea of a balloon by watching the clouds pile one upon the other. According to report, too, they took interest in observing smoke rise from chimneys. But it may well be questioned whether the balloon would have been born at that time had not Etienne read a treatise by Joseph Priestley, *On the Weight of Air*, and shortly thereafter listened to a paper read before the Academy of Lyons by De la Serrière, in which he set forth the theory that objects might be raised in the air by the use of certain gases. Returning to Annonay, Etienne told Joseph his ideas, and in August, 1782, the two began to experiment in ballooning. They tried various gases, including hydrogen, then expensive and little known and called "inflammable air." They soon decided on hot air, created by burning straw and hacked wool. At Avignon, Etienne succeeded in November in getting a tiny balloon of silk, containing forty cubic feet of air, to fly

---

[1] A detailed and interesting account of ballooning in the late 1700's is given by Louis Figuier, *Les merveilles de la science, ou description populaire des inventions modernes* (4 vols., Paris, 1867-1870), II, 423-528. In English, a less detailed but popular narration is presented by W. E. Johns, *Some Milestones in Aviation* (London, 1935), 20-51.

One French writer, Amédée de Bast, *Merveilles du génie de l'homme: découvertes, inventions. Récits historiques, amusants et instructifs* . . . (Paris, 1852), 114, insists on a Chinese origin for the balloon, saying that one was flown at an imperial inauguration in 1306. The story of the Chinese balloon, however, even if true, in no way derogates from the glory of the French inventors, for in eighteenth-century Europe apparently no one had ever heard of it.

about in his room. Back in Annonay, both brothers experimented outdoors with larger balloons. Their second, containing about 600 cubic feet of air, rose 100 to 150 toises (640 to 960 feet). With their third, thirty-five feet in diameter and containing about 23,000 cubic feet, they made their celebrated exhibition of June 5, 1783.[2]

This third balloon, a sphere, was made of wrapping cloth lined with paper.[3] Held erect by a scaffold while being filled, it received the heated air that came from the fire of straw and wool through a large opening at the bottom, fixed in position by a wooden frame sixteen feet square. The demonstration went off smoothly, the balloon rising to a height of more than 1,300 feet and remaining aloft ten minutes before descending lightly on a neighboring hill.

Present in the group of witnesses that day were members of the Estates of Vivarais, then meeting at Annonay. Some of them wrote a report of the experiment to the controller-general of finances at Paris and asked that recognition be given the inventors. This led to an invitation for Joseph to bring the balloon to Versailles to demonstrate it there at the expense of the Academy of Sciences.[4] Arrangements were made for a flight in September.

The Montgolfiers were destined, however, to be anticipated by an earlier flight at Paris. News of what had occurred at Annonay had come to be known to others than members of the Royal Academy of Sciences. The report created a sensation in the capital, and so eager was the Parisian public to witness the marvels of this new invention that a citizen named Faujas de Saint-Fond undertook, with government permission, to raise funds by subscription for a balloon experiment of his own. A popular free-lance professor of physics named J. A. C. Charles

---

[2] "Rapport fait à l'Académie des Sciences, sur la machine aérostatique, de Mrs. Montgolfier . . . ," in *Histoire de l'Académie royale des Sciences, années 1699-1790* (93 vols., Paris, 1702-1797), *1783*, pp. 9-10; Tiberius Cavallo, *The History and Practice of Aerostation* (London, 1785), 43-48; Figuier, *Merveilles*, II, 427.

[3] The paper was tied with string to the cloth. The cloth was in several sections, fastened with buttons. Johns, *Milestones in Aviation*, 23; Cavallo, *Aerostation*, 47.

[4] The expense was found too heavy for the academy, and the government assumed it. *Histoire de l'Académie, 1783*, p. 5 n.

(1746-1823),[5] who lectured at the Louvre and showed eager interest in the new experiment, was placed in charge of overseeing the balloon's construction and operation. Robert Frères, a firm skilled at the manufacture of scientific instruments, was asked to make the balloon.

The balloon, twelve feet in diameter, was of taffeta, covered with a rubber varnish or gum to render it impermeable to hydrogen gas. Knowing that heated air used by the Montgolfier brothers has only twice the lifting power of normal air, Charles decided to employ hydrogen, fourteen times lighter than air.

Filling the balloon with hydrogen was not simple. Charles devised a procedure, however, that worked with fair success. Taking a barrel with two bungholes, he ran from one a leather tube to a valve at the bottom of the balloon. Through the second bunghole he poured iron filings, water, and sulphuric acid; then he plugged the hole. Hydrogen went into the balloon, which was filled for ascension by the scheduled time.[6]

The exhibition took place on August 27, 1783, at the Champs de Mars. A crowd estimated at 300,000—half of Paris—turned out to watch. At the firing of a cannon, the prearranged signal, the highly decorated balloon, called the Globe, was cut loose from its moorings and rose rapidly to a great height. Unfortunately, despite complaint from Charles, Robert Frères had insisted on filling the balloon to its full capacity. This overinflation resulted in a rent in the cloth as the balloon rose to a high altitude where the pressure was less. The escape of gas made the balloon's flight shorter than it otherwise would have been. As it was, it remained aloft forty-five minutes and came to earth in a field at Gonesse, about twelve miles from Paris, greatly frightening the peasants, who according to report at first mistook it for the moon falling and later badly damaged it.

[5] See biographical sketch in *Nouvelle biographie générale depuis les temps les plus reculés jusqu'à nos jours, avec les renseignements bibliographiques et l'indication des sources à consulter*, ed. by J. C. F. Hoefer (46 vols., Paris, 1853-1866), IX, 929-33.

[6] The hydrogen was not free of sulphuric acid fumes and water vapor, and for future performances the gas was run through a barrel of water for cleaning before being sent into the balloon.

The government took note of this fright to peasants by distributing over France copies of a printed bulletin on balloons, telling the people everywhere not to be alarmed if balloons should appear in their neighborhood. The government expressed the hope that the new invention might "some day have applications useful to the needs of society."[7]

Thus there came into being two types of early balloon, that of the Montgolfiers lifted by hot air and that devised by Charles using the inflammable but lighter hydrogen. Both types were destined to receive development and long use. The hot air balloons, called *montgolfières*, soon were fitted with basket and a fire to keep them aloft, and later ballast also was used. The hydrogen balloons, termed *charlières*, were improved by their inventor, who, devised a shutter or valve at the top, opened and closed at will by cords, as well as a method of controlling the valve or tube at the bottom. Charles was the first to use ballast. Later in 1783 an engineering officer named Meusnier suggested the placing of a smaller balloon within the larger balloon as a better means of keeping the latter filled. This form of balloon became common.[8]

On September 19, 1783, Joseph Montgolfier made his exhibition flight before the court and members of the Royal Academy of Sciences at Versailles. It had been set for September 12, but a rainstorm which came at the time for ascension cut short the flight and ruined the balloon. Montgolfier constructed a new one within the amazingly brief period of four days. This balloon, made of cotton cloth reinforced with paper, was beautifully ornamented with blue and gold painting. A huge platform, in the middle of which was a circular opening, was built to hold it. Fastened to this balloon was a basket in which a sheep, a rooster, and a duck were sent aloft, the first aerial passengers. They came down unhurt save for the rooster, which had been kicked on one wing by the sheep before the flight. The great crowd of spectators was kept informed of the various stages of preparation by musketry, according to prearranged understanding.[9]

[7] Figuier, *Merveilles*, II, 432.
[8] Augustin Boutaric, *Les grandes inventions françaises* (Paris, 1932), 251-52.
[9] *Histoire de l'Académie, 1783*, pp. 12-14; Cavallo, *Aerostation*, 66-72; Figuier, *Merveilles*, II, 434.

THE FIRST PASSENGER-CARRYING HOT AIR BALLOON

The *montgolfière* flew over Paris and the Seine November 21, 1784.

(Illustration from Louis Figuier, *Les merveilles de la science*, II, 437, courtesy of the Library of Congress)

The practice of sending animals aloft naturally excited the hope that humans too might fly. Joseph Montgolfier contemplated the idea for a time, but finally drew back lest the balloonists be killed or fire be spread to city or fields by the embers that must be carried. On his suggestion, the king issued a declaration forbidding human flights. Later, after much pressure, the monarch yielded to the point of allowing certain prisoners to be carried aloft. Meanwhile a certain Jean François Pilâtre de Rozier had become intoxicated with the ambition to be the first man to fly, and he used much influence at his command to achieve that end. He succeeded in having the royal declaration amended so that he and the Marquis d'Arlandes might go aloft in a hot air balloon November 21, 1783, in the Jardins de la Muette. The balloon on this flight crossed the city.[10]

These successive developments in ballooning fanned public interest, and when on December 1 Charles and one of the two Robert brothers made their first hydrogen balloon flight, a crowd reckoned at half Paris witnessed the event. Spectators enjoying the favored positions closest to the balloon while it was being filled were charged the fancy sum of four louis d'or (approximately eighteen dollars); others, three livres (sixty cents). At the appointed time the beautiful red and white silk balloon with network of rope and suspended car rose gracefully and remained aloft with its passengers three and a half hours, depositing them near Nesle, some twenty-two miles from Paris. During the flight the balloon drifted completely beyond the sight of the vast concourse in Paris, adding to the thrills of the day.[11]

Unfortunately Charles took no further part in ballooning. The Montgolfiers, too, made no more contributions of significance,

[10] *Histoire de l'Académie, 1783*, p. 17. Pilâtre de Rozier had gone aloft in a captive balloon at Paris on October 15, the anchoring rope being set at different lengths for his several ascensions that day. *Ibid.*, 15-16; Cavallo, *Aerostation*, 74-76.

[11] Figuier, *Merveilles*, II, 442-46; Benjamin Franklin, *The Works of Benjamin Franklin*, ed. by John Bigelow (12 vols., New York and London, 1904), X, 207, 210-15; L. S. Mercier, *Tableau de Paris* (new ed., 12 vols., Amsterdam, 1782-1788), X, 160-64. After the balloon came to rest, Charles insisted on going aloft a second time, alone, that afternoon, taking a barometer and a thermometer and making observations and notes during the half hour of his flight. He later wrote a detailed account of the afternoon's experience, which Figuier, *Merveilles*, II, 444-47, quotes in full.

although they still displayed some interest, as shown by a government advance or gift to them in 1784 for the construction of a balloon.[12] New names displaced theirs in the rapid developments of aviation, but these first inventors were not forgotten. The king rewarded Charles in late 1783 with a pension of 2,000 livres. Letters of nobility were given the father of the Montgolfiers, and Joseph was decorated with the cord of Saint-Michel. The Academy of Sciences awarded the two brothers a special prize of 600 livres. Pensions of 1,000 livres each were granted Joseph Montgolfier, Robert, and Pilâtre de Rozier. The Marquis d'Arlandes was made a colonel, and some months later he was decorated with the cross of commander of the Order of Saint-Louis. Robert, Pilâtre, D'Arlandes, and Charles were named associates of the Royal Academy of Sciences at the meeting of that body December 9, 1783, and the government struck (or at least made plans to strike) a medal bearing the names of Montgolfier and Charles.[13]

The popularity of these airmen, some of them intrepid, was immense. The Montgolfiers headed the list. Fashions of the day were called after them. "Suits were tailored *à la Montgolfière;* hats, parasols, canes were *à la Montgolfière;* society women adopted coiffures *à la Montgolfière;* and the tables *(tréteaux)* at the fair, as well as the tribunes of the learned societies of the little cities in France, sang in every manner the praises of the two physicists of Annonay." Franklin wrote to a friend in December, 1783: "We think of nothing here at present but of flying; the balloons engross all conversation." Tiny balloons filled with hydrogen were sold in large numbers as curiosities.[14]

[12] Maurice Tourneaux, *Bibliographie de l'histoire de Paris pendant la Révolution française* (5 vols., Paris, 1890-1913), III, 183; *Archives parlementaires de 1787 à 1789. Recueil complet des débats législatifs & politiques des chambres françaises,* ed. by J. Madival and E. Laurent (1st ser., 2d ed., 81 vols., Paris, 1879-1913), XIII, 182. The latter gives the date as 1786, but 1784 seems more plausible.

[13] *Archives parlementaires*, XIII, 566; ibid., XIV, 242; Bast, *Merveilles du génie de l'homme,* 122; *Histoire de l'Académie, 1783,* p. 23; Figuier, *Merveilles,* II, 447. It turned out later, under the Empire, that Joseph Montgolfier's widow came to need and wrote a distressing letter to Napoleon, who responded with a pension of 1,200 francs. "Une pétition de la veuve de Joseph Montgolfier à Napoleon Ier," in *Revue historique de la Révolution française,* VIII (1917), 153-54.

[14] Bast, *Merveilles du génie de l'homme,* 120; Franklin, *Works,* X, 215; Cavallo, *Aerostation,* 64-66.

From all parts came demands for balloon exhibitions, and within a year or so flights before great crowds were made at London, Edinburgh, Oxford, Chester, Lyons, Rome—in fact, all European cities of size—and in several American cities, including Le Cap, Santo Domingo, where after the ascension a ball was given and the street where the flight took place named Rue de Ballon. At Edinburgh ascensions were made in March, September, and October, 1784.[15] In many cities there were no doubt several flights. No other invention of the century so caught the fancy of the public.

On June 4, 1784, a year after the first flight, Pilâtre de Rozier and a chemist named Proust soared at Lyons in a *montgolfière* to an altitude of 11,732 feet, where they were on a level with the highest mountains. For several hours they remained in flight as their craft moved over mountain peaks and they looked down on clouds below. At times the earth was hidden from view; at times they had to add fuel to rise above approaching peaks. They got a great thrill from playing leapfrog with the mountaintops.[16]

It was natural that crossing the English Channel by balloon would occur to someone. After a small balloon sent up by an Englishman, William Boys of Sandwich, Kent, with a letter asking that the finder please communicate with him landed in Normandy,[17] a race to be the first man to fly between England and France began. In late 1784 two rival balloonists, François Blanchard and Pilâtre de Rozier, made ready for the crossing. Pilâtre was first to declare his intention. With an assistant he planned to fly from Normandy to England in a government-

[15] Cavallo, *Aerostation*, 109-11; Mercier, *Tableau de Paris*, X, 160-64; Carlos d'Eschevannes, "L'aviation en 1783," in *Nouvelle revue*, 3d ser., XVII (1910), 423-29; Louis Petit de Bachaumont, *Mémoires secrets pour servir à l'histoire de la république des lettres en France, depuis MDCCLXII jusqu'à nos jours; ou, journal d'un observateur* (36 vols., London, 1777-1789), XXV; Dorothy Mackay Quynn and William Rogers Quynn, "Letters of a Maryland Medical Student in Philadelphia and Edinburgh (1782-1784)," in *Maryland Historical Magazine*, XXXI (1936), 202, 210, 211; H. Castonnet des Fosses, *La perte d'une colonie: la révolution de Sante-Domingue* (Paris, 1893), 22; Thomas Baldwin, *Airopaidia: containing the Narrative of a Balloon Excursion from Chester, the Eighth of September, 1785* . . . (Chester, 1786).

[16] Pilâtre's account is reproduced by Figuier, *Merveilles*, II, 459-60.

[17] Johns, *Milestones in Aviation*, 32.

financed balloon. While he made preparations, however, Blanchard crossed by boat to England and from there with an American Loyalist named Dr. Jeffries[18] as companion flew from Dover to Calais, January 1, 1785. The flight was made under excellent flying conditions. The day was cold, but there was a slight wind. At first the balloonists maintained a great height and enjoyed a magnificent view of eastern England and of shipping in the Channel. But after a time the chilly atmosphere caused the balloon to contract and lose altitude. In an effort to counteract this the passengers threw overboard their anchor, their food, and part of their clothing and other belongings. As this seemed to bring no change for the better, Jeffries offered to carry out his contract to jump overboard if the balloon were in danger of falling; but to this Blanchard would not hear. Instead, he insisted that they survive or sink together; and they were about to cut away the basket when suddenly the balloon, evidently warmed by the sun, rose again and carried its passengers safely to the shores of France, coming to rest in the forest of Guines, near Calais. The two balloonists were honored at Calais with a reception, the mayor and citizens presented Blanchard with a gold box, and the balloon was put on display. Louis XVI awarded Blanchard a small pension, which was paid until the Revolution.[19]

Later in 1785 Pilâtre de Rozier and his assistant, after idling many months on the Channel and incurring a government reprimand, as well as the public reproach of cowardice, set out to reverse Blanchard's flight. Unfortunately Pilâtre, who up to now had relied wholly upon *montgolfières,* used on this occasion a hybrid form of balloon, carrying both hot air and hydrogen.[20] Hardly were they well into the air when the hydrogen exploded and balloon and aeronauts fell to the ground. Pilâtre thus became the first fatality in ballooning history.

[18] Jeffries was an American physician who had been trained at Harvard College. His Loyalist sympathies led to his emigration to England. A man of means, he gave financial backing to Blanchard on this trip. According to report he paid £100 to make the passage. *Ibid.,* 33.

[19] *Ibid.,* 32-36.

[20] Figuier gives not only a description but an illustration of this balloon. Its upper part was round, filled with hydrogen, and closed. Subtended was a cylinder filled with hot air. As was his custom, Pilâtre carried a fire in the basket to feed the *montgolfière* portion.

While these developments in ballooning were taking place, certain experiments were being made to develop a parachute. Originally conceived by Leonardo da Vinci, who drew a sketch of one as a means by which a man might safely descend from a great height, the parachute was first used by Sebastien Lenormand, who successfully jumped from a tower in Montpellier and gave the safety device its name. He used an umbrella with a radius of thirty inches, reinforced by cords tied to the ends of the ribs and held in his hands. He first jumped from a one-floor elevation. Later he tried dropping animals of various sizes, and noted a relation between the weight of the animal and the size of the cover needed. He concluded that a parachute forty feet in diameter would well support a man for long descents, provided that the man and parachute did not weigh over 200 pounds. He himself jumped with an umbrella from the tower of the Observatory in late December, 1783. The Estates of Montpellier were meeting in the city at the time, and their members as well as one of the Montgolfiers witnessed the event. Thus the parachute—a development of the umbrella—was originated in the same year as the balloon.

Shortly afterward, Blanchard dropped from his balloon over Paris animals in baskets fastened to a large umbrella, and they came to earth safely.[21] On one occasion, according to report, he dropped his dog from a height of 6,500 feet. Neither Blanchard nor Lenormand, however, attempted to jump from a balloon himself, and the honor of being the pioneer at this feat went to A. J. Garnerin, a former pupil of Charles, who jumped from a height of 3,000 feet over Paris before a host of gaping spectators October 22, 1797. He later made exhibition jumps throughout Europe, as did also his wife and his niece Elise.[22] Thus from an

---

[21] Figuier, *Merveilles*, II, 521-23; *Biographie universel, ancienne et moderne; ou, histoire, par ordre alphabétique de la vie publique et privée de tous les hommes qui se sont fait remarquer par leurs écrits, leurs actions, leurs talents, leurs vertus ou leurs crimes*, ed. by L. G. Michaud and J. F. Michaud (85 vols., Paris, 1811-1862), LVIII, 341.

[22] Figuier, *Merveilles*, II, 522-23; Johns, *Milestones in Aviation*, 46-51; A. Rambaud, "Les sciences pendant la Révolution et l'Empire," in *Révolution française*, XIII (1887), 117. Garnerin's first use of the parachute was in 1793, when he attempted to escape from an Austrian prison by means of one. It failed him, and he broke his leg in the fall. Jacques Godechot, "L'aérostation

early date women assumed a role in aviation. The parachute used by Garnerin was a giant parasol of canvas, to which a basket was attached by numerous ropes. Other parachutes that came soon into use were more commonly pieces of silk sewn together in the form of an umbrella bound at the top to a circular piece of wood. To this wooden ring were attached a great number of ropes, which fell over the roof of the parachute and held in suspension the basket or car for the aeronaut. A small opening in the center of the wooden disk permitted the passage of air to prevent damage to the parachute.[23]

From the outset the idea of using balloons for military purposes presented itself. Gerond de Villette, passenger with Pilâtre de Rozier on his first ascension (by captive balloon), October 15, 1783, wrote to the *Journal de Paris:* "At once I was convinced that this machine, although somewhat expensive, might be very useful in war to enable one to discover the position of the enemy, his maneuvers and his marches; and to announce these by signals to one's own army. I believe that at sea it is equally possible to make use of this machine."[24] The government took no interest, however, until 1793. In October of that year a physicist named Bonnemain announced to the National Convention that he had invented "a machine by which death and destruction can be carried from the air to enemy battalions and squadrons." He offered it to the French government, then at war with Austria, Prussia, Great Britain, Spain, and the Netherlands. The government appointed a commission of three, consisting of the scientists Count de Fourcroy and Baron Guyton de Morveau and the deputy Moreau, to investigate its merits and at the same time placed 6,000 livres at the disposal of the minister of the interior to enable Bonnemain to carry his experiments further. Thus far, according to Moreau, the "machine" had conquered the incredulity of those who had examined it and appeared capable of fulfilling the prom-

---

militaire sous le Directoire," in *Annales historiques de la Révolution française,* VIII (1931), 221. Godechot places the height for the Paris jump at "around 1,000 meters."

[23] Johns, *Milestones in Aviation,* 47-49, quotes at length a late eighteenth-century description of a parachute and its operation. The changes from that day to this have been relatively small.
[24] Quoted *ibid.,* 37.

THE FIRST PARACHUTE JUMP

Jacques Garnerin descended in the basket of his balloon from a height of 1,000 meters October 22, 1797, over the Seine near Paris.

(Illustration from Figuier, *Merveilles*, II, 525, courtesy of the Library of Congress)

ises of the inventor. Its details, he stated, could not be discussed in public as it was a secret weapon.[25]

What became of Bonnemain's invention is not related, but Guyton de Morveau became convinced that balloons might serve a useful function in war. About this time he was placed on a commission with Gaspard Monge, Count de Berthollet, L. N. M. Carnot, Fourcroy, J. J. L. F. de Lalande, and A. L. Lavoisier to utilize recent inventions for the best interests of the state. He set forth his viewpoint concerning the balloon and persuaded the commission to recommend it to the committee of public safety.[26] This body agreed to the experiment, provided that no sulphuric acid be used for making hydrogen, since it was badly needed in making gunpowder. The difficulty was easily solved. Lavoisier had recently discovered a means of making hydrogen by decomposing water with red-hot iron filings. The oxygen in the water united with the iron, leaving the hydrogen free. Experiments showed that hydrogen could be made on a large scale by this method, but that a furnace would be necessary. Assured of this, the committee of public safety gave orders to proceed with military ballooning.[27]

J. M. J. Coutelle (1748-1835), a former student of Charles in physics and a friend of Guyton de Morveau, was placed in charge; Jacques Conté, another young scientist, was made second in command. Other officers in the Ballooning Corps were Lhomond, a young physicist, and Delaunay, a brickmason chosen to construct furnaces. A company of fifty or sixty men was placed in their charge. They constructed a balloon of gummed silk, thirty feet in diameter, which they named the *Entreprenant,* destined

---

[25] *Archives parlementaires,* LXXVI, 309; *Réimpression de l'ancien Moniteur* [*universel*], *seule histoire authentique et inaltérée de la Révolution française depuis la réunion des Etats-généraux jusqu'au Consulat (mai 1789 - novembre 1799), avec des notes explicatives,* ed. by A. Ray (32 vols., Paris, 1858-1863), XVIII, 94 (hereafter, *Moniteur universel*).

[26] Figuier, *Merveilles,* II, 490; Bast, *Merveilles du génie de l'homme,* 125.

[27] Godechot, "Aérostation militaire," 217-18, attributes this method of obtaining hydrogen to Conté; actually, it had been devised in 1783 by Lavoisier and Meusnier. See A. Wolf, *A History of Science, Technology, and Philosophy in the Eighteenth Century* (New York, 1939), 371. According to Wilfrid de Fonvielle, *Histoire de la navigation aérienne* (Paris, 1907), 42, its use on this occasion was suggested by Lavoisier.

to have a celebrated record. So well was this balloon constructed that it was able to retain its load of hydrogen gas for two or three months without any appreciable loss, and thus to be kept in readiness.[28] The soldiers mocked, and even J. B. Jourdan in charge of the army was derisive toward the "toy," but on orders from Paris it was put to the test. It was used for observation in the sieges of Condé (1793), Maubeuge (1794), and Charleroi (1784); in the battles of Fleurus (1794) and Gosselins (1794); and later in the campaign along the Rhine (1795).[29]

In each instance two balloonist officers went aloft in a balloon held captive with two ropes by sixteen men. Messages to the ground crew were communicated by the use of red, yellow, and green flags some eighteen inches square; messages to the general were dropped in bags weighed down with ballast and marked by a pennant or streamer. No one might handle these last save one of the Ballooning Corps officers. The balloon made a great impression on the Austrians, who on one occasion attempted with near success to shoot it down, but oddly enough did not attempt to imitate it.

The extent of the services rendered by the balloon has been disputed. Immediately after the battle of Fleurus, Jourdan spoke highly of the work of the *Entreprenant,* as also did Fourcroy who was present; later in 1799 Jourdan gave a disparaging report to the Directory on the value of military balloons. On the other hand, Baron de Selle de Beauchamp, at first a private and later a lieutenant in the Ballooning Corps, related in his *Mémoires* that the service of the early balloons was great, particularly at Fleurus, where the French won a victory enabling them to drive the Austrians from French soil and to invade Belgium.[30]

---

[28] Figuier, *Merveilles,* II, 495. Conté later devised a closer-woven fabric for balloons and also a varnish that enabled them to remain inflated for a period of four months. Godechot, "Aérostation militaire," 217.

[29] At Valenciennes (1793) a French balloon was captured by the Allies, and with it a pigeon carrying dispatches. The enemy indulged their humor by eating the pigeon and by firing the balloon back into the town from a cannon. Ramsay Weston Phipps, *The Armies of the First French Republic and the Rise of the Marshals of Napoleon I,* ed. by C. F. Phipps and Elizabeth Sandars (5 vols., London and New York, 1926-1939), I, 184.

[30] Figuier, *Merveilles,* II, 501-503; R. R. Palmer, *Twelve Who Ruled: The Committee of Public Safety during the Terror* (Princeton, 1941), 350-57; G. Pouget, "Les sciences pendant la Terreur," in *Révolution française,* XXX (1896), 339.

Meanwhile the Convention organized a second company of balloonists and created (October 31, 1795) the National Ballooning School at Meudon, where sixty cadets were to be trained in physics, chemistry, mechanics, and geography for future service as balloonist officers. Conté, who had distinguished himself in balloonist operations at the front, was named director of the school. Twenty-two other balloons, bearing such names as *Céleste, Intrépide, Hercule, Martial, Aigle,* and *Lion,* were sent to the armies of the North.[31]

In 1798 Napoleon, when he set out for Egypt, took with him one of the ballooning companies that had served against the Austrians, with Conté and Coutelle in charge. Unfortunately the balloons, on the warship *Orient,* were captured at Alexandria by the British fleet, while most of the ballooning equipment, loaded on the *Patriot,* went down with that vessel when it ran aground and sank near Aboukir. Conté, skillful improviser that he was, constructed three *montgolfières,* with which exhibitions were made at Cairo, dazzling the Egyptians. The balloonist officers, however, accustomed to the use of hydrogen, were unable to obtain in Egypt the materials for making the gas. The Balloonist Corps was therefore of no military value on the expedition. Napoleon from this time forth had no interest in balloons. The military critic B. H. Liddell Hart has remarked in a touch of irony that "if Napoleon had not stifled this promising infant, on its twenty-first birthday in June, 1815, it might have saved him from disaster" at Waterloo—almost precisely the spot where the balloon was first effectively used in warfare.[32]

The chief handicap of the early balloons was the inability to guide them. Numerous attempts were made to overcome this deficiency. On his first flight over Paris, March 2, 1784, Blanchard attached wings and a rudder to his balloon to make it dirigible.

[31] Figuier, *Merveilles,* II, 504; Godechot, "Aérostation militaire," 215-16. The French used balloons in their operations along the Rhine and in Austrian territory in 1796, losing certain of them to the enemy at Würzburg. John Goldstrom, *A Narrative History of Aviation* (New York, 1930), 16.

[32] B. H. Liddell Hart, "Would Another War End Civilization?" in *Harper's Magazine,* CLXX (1935), 316; Godechot, "Aérostation militaire," 224. According to Godechot, Napoleon had requested the Directory for a ballooning company for his first Italian campaign (1796), but there is doubt whether it was used.

The wings, however, were rudely broken in anger by a student from the Ecole Militaire who rushed forward just as the balloon was ready to rise and demanded, without success, that Blanchard allow him to go aloft with the party.[33] The youth also wounded Dom Pech, a Benedictine monk and physicist who was to have accompanied Blanchard on the flight. Blanchard went aloft alone, and always insisted that he successfully navigated the balloon by means of the rudder; but this feat was attributed by scientists to the air currents in which Blanchard found himself.[34] Certainly he was unable to repeat his feat, nor was anyone else able to do so.

In a remarkable paper read before the Academy of Sciences in December, 1783, and published as a book in 1784 under the title *Atlas des dessins relatifs à un projet de machine aérostatique*, J. B. M. Meusnier (1754-1793), a young lieutenant of engineers, proposed construction of a huge dirigible balloon, ellipsoidal in shape and 260 feet long, "to be propelled by three [revolving] air-screws worked by hand." His plan was not tested, but according to Professor A. Wolf, "it may be regarded as the starting-point of an evolutionary series of dirigible balloons and airships which were constructed, or merely designed, during the course of the nineteenth century."[35] Meusnier also conceived of using within balloons smaller balloons filled with air to control or cushion their rise and fall. He was a brilliant young scientist, a graduate of the engineering Ecole de Mezières, where he won the admiration of Monge, who "considered him the most extraordinary mind he had ever known." Unfortunately for France, he was killed at the siege of Mainz (1793) at the age of thirty-eight.

Other proponents of the dirigible balloon appeared before the turn of the century. In February, 1789, Baron de Scott, a captain of dragoons, sought police permission to raise a subscription for "a new aerostatic machine, dirigible at will."[36] On May 2,

---

[33] Several instances of would-be aerial hitchhikers are found in the early annals of ballooning. This one and another at Lyons are related by Figuier, *Merveilles*, II, 451-53, 457.

[34] *Biographie universelle*, LVIII, 341.

[35] Wolf, *History of Science*, 581; Maurice Caullery, *French Science and Its Principal Discoveries Since the Seventeenth Century* (New York, 1934), 64-66.

[36] Alexandre Tuetey (ed.), *Répertoire générale des sources manuscrites de l'histoire de Paris pendant la Révolution française* (11 vols., Paris, 1890-1914), III, 539.

1792, a retired army officer named Delacourcière appeared before the Legislative Assembly and reported that after eight years of experimentation he had found a way of directing balloons and that he could construct balloons capable of transporting two hundred men, equipped to the extent of forty-eight pounds with guns, ammunition, and food, for service in all parts of the world.[37] History has left no record of such fantastic claims being tested. That they could actually be put into effect remained for Zeppelin and other Germans of the late nineteenth century to demonstrate.

Another eighteenth-century advocate of the dirigible came forward in Paul Lamanon. At a meeting of the National Convention August 4, 1793, a petition from him was read in which he told of possessing a means of directing balloons and observing the movements of enemy armies. He claimed to have covered the distance of four miles with his balloon in twenty-three minutes and to be able to direct it as he wished. He bade the French not to despair over the recent distressing news of the fall of Valenciennes, for he had a means—a secret weapon—for saving the country. He requested only that the Convention place at his disposal a sum of money necessary to purchase materials and to pay the workers needed to assist him. The matter was referred to the committee of public instruction, and no further mention of it is found in the records.[38]

In the same year two citizens, Marre and Desquimare, sent to the Convention a long description of a dirigible balloon which they claimed to have invented and which they thought might be useful to the French government in its war. Its chief feature was two huge wings, each consisting of a series of parallel planes, like feathers in a bird's wing, which might be worked up and down by cords controlled by the two balloonists from a central basket or carriage. The "feathers" were of great size. The alleged inventors, however, did not claim to have tried out their fantastic device.[39]

[37] *Archives parlementaires*, XLIII, 5.
[38] *Procès-verbaux du comité d'instruction publique de la Convention nationale*, ed. by James Guillaume (6 vols., Paris, 1891-1907), II, 284-85. The petitioner was evidently a son of Robert Paul Lamanon (1752-1787), naturalist and traveler killed by natives on the island of Maouna. See *Biographie universelle*, XXIII, 255-56.
[39] Gustave Vauthier, "Un essai de ballon dirigeable en 1793," in *Revue historique de la Révolution française*, VI (1915), 305-11.

Interestingly enough, more than one eighteenth-century visionary labored at the idea of flying by aeroplane. From the days of the legendary Icarus men have dreamed and attempted to fly by means of mechanical wings. The idea accordingly was not new; nevertheless the French attempts of the eighteenth century have their interest. In 1772 one Desforges, canon of Etampes, announced in the public journals that he would attempt to fly a complicated machine equipped with movable wings, which he designated a "flying carriage" (*cabriolet volant*). The canon was as good as his word and tried the flight as announced, but his machine did not measure up to his expectations. He fell and was injured. Undismayed, he made other attempts with unusual devices, and in a letter of October 24 that year he wrote: "I believe that one might fly in the air if he could find a means of force almost infinite." This bizarre man got his interest in flying, it seems, from his study of nightingales.[40]

The Marquis de Bacqueville likewise attempted a flight across the Seine from a window on the Rue des Saints-Pères by means of wings, attached to his shoulders, which he was able to move by a mechanical contrivance. Like the reckless canon of Etampes, he fell, landing on a laundry boat and suffering a broken thigh.[41]

Equally as visionary was a heavier-than-air machine at which Blanchard worked for some years prior to 1783. It was shaped like a bird, with body of light but solid wood, convex above and below, and pointed fore and aft. There were six wings, each ten feet square, which the passenger was able to work rapidly in a manner like a bird in flight. The machine, however, lacked motive power. Long and strenuously Blanchard toiled at perfecting this invention, and he was reportedly on the point of

---

[40] Desforges also became interested in the love life of nightingales. He decided that the clergy also needed love life and wrote a book advocating their marriage, for which he was thrown into the Bastille and his book burned. According to Grimm, the daring canon carried through his view in this matter as in others and married "a Christian girl"; but the fact is not certain. See the account in *Nouvelle biographie générale*, XXX, 807-808.

[41] Paul Lacroix, *XVIIIme siècle: lettres, sciences et arts, France 1700-1789* (Paris, 1878), 65-66. Whereas Lacroix indicates that this flight occurred after the experience of Desforges, Goldstrom, *History of Aviation*, 12-13, places it in 1740.

going abroad in search of financial aid not forthcoming in France when the Montgolfiers attained success with the balloon. Thenceforth Blanchard deserted his machine for balloons.[42]

Not only the invention but also the improvement of the balloon was exclusively a French affair. But what benefits from it accrued to France? What profit could be made of it, save at exhibitions, and how could developments be kept secret? It was not the type of invention to bring financial returns. A pessimist in November, 1783, is reported to have visualized this and made the cynical remark: "What good can come from balloons?" Benjamin Franklin, who saw Charles' ascension in August, is said to have retorted: "What good can come from a baby just born?"[43]

Franklin, an enthusiast for the balloon, wrote in his report of August 30, 1783, to Sir Joseph Banks of the Royal Society of London that "possibly it may pave the way for some discoveries in natural philosophy of which at present we have no conception." He thought that it might aid the cause of world peace by rendering armies and navies obsolete. It might inaugurate a new mode of travel for men. It might provide a new source of refrigeration, for the preservation of game and the obtaining of ice when needed. Other uses, too, might come from it. "This experiment," he said, "is by no means a trifling one. It may be attended with important consequences that no one can foresee."[44]

---

[42] *Biographie universelle*, LVIII, 337-38. Figuier, *Merveilles*, II, 517-19, however, attributes this device to an Italian, and says that Blanchard constructed a flying boat with oars and a second machine with two wings and a pulley. He exhibited his machines, but had the good sense not to jump from buildings in them.

[43] Figuier, *Merveilles*, II, 438.

[44] Franklin, *Works*, X, 158, 161, 208, 215, 267.

CHAPTER II

Steam Transportation

PERHAPS NO MORE SIGNIFICANT DEVELOPMENT WAS MADE during the eighteenth century than in the steam engine. Invented in the late 1600's by Denis Papin, it shortly underwent improvements by Thomas Savery and Thomas Newcomen of England in 1698 and around 1711, and in the 1760's and 1770's it was perfected further by James Watt, the Scot, who gave it closer fittings, a separate cylinder, and double motion of the piston, thereby greatly increasing its power and speed. All of these men saw the opportunity for practical application of this new source of motive power, destined in the nineteenth century to transform the world, and tried to apply it, Papin to a steamboat, Savery and Newcomen to the drainage of mines, and Watt to industry. Their story is well known. They and others realized that a great source of energy had been created which could be of enormous advantage to man if properly applied, although only Watt lived to see his dreams in part realized. While the French lagged behind the British in the 1700's in the perfection of the steam engine itself, they went farther than the British in their vision of its application, or rather in the attainment of their vision.

On July 15, 1783, six weeks after the first exhibition of the balloon at Annonay and thirty miles distant, a crowd assembled on the banks of the Saône at Lyons saw the public trial of the first successful steamboat.[1] The inventor was the comparatively young

---

[1] Charles Ballot, *L'introduction du machinisme dans l'industrie française*, ed. by Claude Gével (Paris and Lille, 1923), 390 n. 1, a normally reliable French authority, states that the first steamboat was sailed in America in 1763 by William Henry; but the article on Henry in the *Dictionary of American Biography*, ed. by Allen Johnson, Dumas Malone, and Harris E. Starr (21 vols. and index, New York, 1928-1944), VIII, 561, says that his boat was not successful (hereafter, *D.A.B.*).

John Fitch, who in 1787 designed a steamboat that plied the waters of the Delaware River at Philadelphia success-

Marquis de Jouffroy d'Abbans (1751-1832), a native of Franche-Comté. Success in this instance had come not at once, but only after a century of thought and experiment by men of various nations, France above all.

The first reportedly to conceive of the steamboat was Denis Papin (1647-1714), Huguenot refugee and wanderer, by turns a resident of England, Italy, and Germany.[2] Around 1670 he had been assistant to Christiaan Huygens, the great Dutch scientist, who introduced him to experiments with the air pump and vacuums. From these beginnings he developed an interest in the pressure created by steam. As early as 1697 he had designed a method of propelling a boat by steam, and presented a model to the Royal Society of London, of which he was a member. Later in the same year he left England to accept a professorship of mathematics at the University of Marburg, in Germany, and gave no further thought to the steamboat until 1705, when he received from Baron von Leibniz, traveling in England, a drawing of the steam engine of Thomas Savery. From then until 1707, the idea of a steamboat dominated Papin's mind, and he acquired a boat at Cassel on the Fulda with which he hoped to go downstream to Bremen and then to England, for experimentation there with steam power.

In the mid-nineteenth century some correspondence of Leibniz was discovered and published, in which were certain letters written by Papin in 1707 and also a letter by Zeuner, bailiff of Münden, telling of Papin's trip by boat down the Fulda to Münden, of his unsuccessful application to the elector of Hanover for permission to take the boat from Münden to Bremen on the Weser (the boatmen protested that they had monopolistic rights on the Weser), and finally of damage to his craft by the boatmen.[3]

---

fully with members of the Continental Congress as witnesses, began his experiments in 1785. *D.A.B.*, VI, 425; Wolf, *History of Science*, 569-70. Also in 1787 success with a steamboat on the Potomac River near Shepherdstown, (West) Virginia, was attained by James Rumsey of Maryland. *D.A.B.*, XVI, 223.

[2] Ballot, *Machinisme*, 389. The report of Blasco de Giray moving a boat by "boiling water" in the port of Barcelona July 17, 1543, is fantastic. The steam engine had not then been invented. For the story, see Figuier, *Merveilles*, I, 151.

[3] Figuier, *Merveilles*, I, 58-59, reprints the letters.

Reading this correspondence, one gains the impression that the boat traveled from Cassel to Münden under steam, and that Papin was the first to demonstrate the steamboat. This conclusion, however, has been shown to be erroneous by certain later writers who have set forth evidence that Papin used men to handle the oars in making this voyage. He planned to install a steam engine when he reached England.[4]

From Papin's letter of September 15, in which he told Leibniz that his boat had succeeded as well as he had hoped, it appears that he had installed a mechanism for turning the oars and that only a steam engine was absent.[5] This in itself was not a novelty. The Romans, as early as the first Punic War, used oxen aboard ships and rafts to turn a mechanism controlling paddle wheels, and troops were sent on transports to Sicily in this fashion.[6] In 1732 the Comte (later Maréchal) de Saxe submitted to the Academy of Sciences two plans for moving boats by means of sidewheels and horsepower. Two horses on deck were to work a treadmill in each case. The drawings reveal the idea to have been practical, but there is no account of its having been put into operation.[7]

After Papin, two Englishmen, J. Dickens in 1724 and Jonathan Hulls in 1737, gave their attention to the idea of a steamboat, but neither met with success. In 1753 the Academy of Sciences at Paris offered a prize for the best paper on "Means of Supplying the Action of Wind for Moving Vessels." Leonhard Euler, the Abbé Gautier, and Daniel Bernoulli were among those submitting papers. Bernoulli won. He is said to have proposed "either a reaction jet of water or oars at the stern and sides of a vessel, driven by steam or animals on a treadmill." Gautier,[8] whose

---

[4] Ballot, *Machinisme*, 390 n. 1.

[5] Apparently it was this mechanism to which the boatmen objected and which they damaged in their attack.

[6] Figuier, *Merveilles*, I, 163. Some ancient Roman coins show Roman warships moved by three pairs of oxen.

[7] *Machines et inventions approuvées par l'Académie royale des Sciences, depuis son établissement jusqu'à présent: avec leur description*, ed. by Gallon (7 vols., Paris, 1735-1777), VI, 37-44.

[8] Eric Hodgins and F. Alexander Magoun, *Behemoth: The Story of Power* (Garden City, N.Y., 1932), 173; Figuier, *Merveilles*, I, 152-53. The writers of the former book say that the Abbé Gautier's paper was submitted not in this contest but to the Académie des Sciences et Belles-Lettres at Nancy. Inasmuch as it was published shortly afterward, in the third volume of some papers by the Abbé, and bore the precise title for papers entered in the con-

paper was published later, was more wholeheartedly in favor of the steamboat, however, than Bernoulli.[9] A few years later the Swiss clergyman Genevois published at London a pamphlet[10] in which he advocated a steamboat equipped with an engine like Thomas Newcomen's but driven by gunpowder.

Thus during the century intermittent attention had been given to the idea of moving boats by new mechanical means, particularly by steam. Beginning with the 1770's this interest in steam as a source of nautical motive power greatly increased.

In 1771 or early 1772 the former artillery officer Chevalier Claude François Joseph d'Auxiron (1731-1778), a native of Besançon, conceived of setting up an inland navigation company, operating steamboats on the larger French rivers—the Seine, Loire, Garonne, and Rhône—and possibly attempting oceanic trade. For some years (1763-1771) he had worked with the idea of installing a system to supply Paris with water by means of the steam engine, and had applied to the government for the privilege, but the Academy of Sciences to which the government referred the matter adopted the advice of Lavoisier in vetoing the idea. No sooner was this application rejected than D'Auxiron transferred his attention to steamboat navigation. He aroused the interest of his lifelong friend and fellow army officer, the Chevalier de Follenay, and the two agreed to work together. D'Auxiron was to have charge of the mechanical phases of the undertaking; Follenay, of the financial. They organized a company with three fellow Franc-Comtois citizens who promised to supply the needed capital. D'Auxiron then applied to the government for monopolistic privileges for his company. This was not quite customary procedure, as it was proper first to produce the invention and then to ask for rights of exploitation. But so confident of success was D'Auxiron that he did not think it necessary or prudent to wait.

---

test: "Manière de suppléer à l'action du vent sur les vaisseaux," it would appear that the paper was submitted as claimed by Figuier.

[9] The Abbé proposed in this paper not only to use steam to do the work of oars, but also "to raise anchors, to cook foods, and to renew the air." An account of the Abbé is in *Nouvelle biographie générale*, XIX, 722-25.

[10] Entitled *Quelques découvertes pour l'amélioration de la navigation*.

Bertin, the minister to whom he appealed, took a different viewpoint, however. He replied that monopolistic rights for a period of fifteen years would be granted "if, when you have placed this method in practice, it is found by the Academy of Sciences truly useful to navigation"—in short, when D'Auxiron had submitted his invention and it had passed the test.

At once D'Auxiron started construction of his boat, but he ran into hostility from the corporations of boatmen and transport workers and was forced to apply for military protection. When finally on April 21, 1771, the boat was completed, Périer, a celebrated machinist, paid it a visit. The menacing attitude of the boatmen and others led D'Auxiron to move the boat from the Isle des Cygnes to a point near Meudon. Even there, as time revealed, it was not safe. One night in September, before a test at sailing, the boat was sunk. A counterweight of 130 pounds used in controlling the piston had been dropped by Captain Bellery through the boat's bottom and by the next morning the craft had disappeared. Malice was suspected.[11]

D'Auxiron was dejected and his financial backers angry. They charged him with foul play and refused to supply further capital, even for refloating the vessel, until he brought suit in court and won the decision.[12] The court specified that the company should provide 15,000 livres toward raising the boat. The decision, however, appears to have been an empty one, for much bad feeling had developed and the company soon went to pieces.[13]

It was shortly after this that the Marquis de Jouffroy d'Abbans made his appearance. Like D'Auxiron and Follenay, he was a native of Franche-Comté and had served as an army officer. For fighting a duel with a superior officer he had been sentenced through a *lettre de cachet* to a prison on the Isle Sainte-Marguerite, near Cannes. There in prison, according to report, he watched

[11] Ballot, *Machinisme*, 390-92; Figuier, *Merveilles*, I, 157-59. Why D'Auxiron delayed until September to make a test, when the boat supposedly was ready in April, is not explained.

[12] His backers claimed that he realized the boat's incapacity and scuttled it.

[13] Already before its sinking the boat and its steam engine had cost 15,200 livres, in addition to 14,000 livres in wages owed the workmen. The court ordered that the arrears in wages be paid.

the galleys plying the sea and contemplated the possibility of propelling ships by some mechanical means. Released in 1775, he visited Paris. There one of the first curiosities to catch his attention was the celebrated machine at Chaillot, installed by the firm Périer Frères. A fire pump powered by a Watt steam engine, it was one of the attractions of the capital, visited by crowds of spectators. Jouffroy studied it intently, obtaining from Périer Frères permission for a close examination. The idea of using a Watt engine for propelling a boat came to his mind.

He joined a new company that D'Auxiron and Follenay were forming in Paris at this time, whose central figure was Ducrest, brother of Madame de Genlis and a man of wide acquaintances and court favor. Another member was Périer. He and Jouffroy differed over the boat's construction, Jouffroy insisting that more power must be given the side wheels than Périer thought necessary. When the company naturally supported Périer, a man of long experience and great reputation as a mechanical engineer, Jouffroy withdrew and returned to his native Franche-Comté to work alone.[14]

Hiring a coppersmith in the little town of Baume-les-Dames, on the Doubs, he set to work building and equipping a small boat forty feet long and ten feet wide. In its middle he placed a Watt steam engine; on each of its sides, a paddle wheel. The paddles were web-footed, after a suggestion by the Swiss pastor Genevois. They moved only when the piston was descending; a counterweight, which rose and fell in the water, carried the piston back.[15]

When this boat proved unsuccessful, Jouffroy in 1780 began to construct a second and larger vessel on the Saône at Lyons, the ship that was to bring him fame. It was 141 feet long, 14.7 feet wide, and equipped with side wheels 14 feet in diameter, having paddles (straight and not web-footed) 6 feet long which dipped two feet into the water. The boat itself drew three feet of water, weighed 27,000 pounds, and had a carrying capacity of 300,000 pounds. Instead of using again his Watt steam engine,

[14] Incidentally, Périer's boat did not work, and the company dissolved. Figuier, *Merveilles*, I, 162.
[15] *Ibid.*, 162-64.

he used another, reputedly more powerful, with a piston 5 feet long and 21 inches in diameter, built by the firm Frères Jean of Lyons.[16]

With this boat Jouffroy made his public exhibition July 15, 1783. Before a cheering crowd of thousands of spectators, including members of the Academy of Lyons, he successfully navigated the boat under the power of steam up the Saône to the Isle Barbe. According to Figuier, this was not the first display of the boat's capacity, for it had several times previously been taken from its mooring point at Lyons to the Isle Barbe under steam, in the presence of large crowds, on trips not regarded as public exhibitions.[17]

After the formal demonstration of his craft, Jouffroy set about organizing a company to operate steamboats on the Saône, the Rhône, and other French rivers for the transportation of merchandise. Among the members of this company were Follenay and a brother of D'Auxiron. Application was made to Charles Alexandre de Calonne, controller general of finance, for privilege. Follenay went to Paris to press the matter. De Calonne, following customary procedure, referred the question to the Academy of Sciences, which in turn formed a committee to investigate. After some time the academy replied that before the alleged invention could be given official recognition it must be brought to Paris and there demonstrated before its committee. De Calonne wrote Jouffroy to this effect January 31, 1784, saying: "I return you, Monsieur, the attestation of success of the fire pump [that is, steam engine] at Lyons, which you propose to substitute for horses for river navigation, and also some other pieces that you have addressed to me in support of your request for exclusive privilege with machines of this kind for a certain number of years. It turns out that the demonstration made at Lyons does not sufficiently satisfy the required conditions; but if, by means of the fire pump, you succeed in taking a boat up the Seine the distance of several leagues, proven or certified in such a manner as to leave no doubt on the value of your procedures, you may

---

[16] *Ibid.*, 163, 165-66; Ballot, *Machinisme*, 394. It seems odd that a Watt steam engine should have been discarded, for the Watt engines were considered the best in the eighteenth century, and like all English machinery were obtained in France with great difficulty.

[17] Figuier, *Merveilles*, I, 166.

THE FIRST SUCCESSFUL STEAMBOAT

This boat, built by the Marquis de Jouffroy d'Abbans, moved several hundred yards against the current of the Saône at Lyons July 15, 1783.

(Illustration from Figuier, *Merveilles*, I, 165, courtesy of the Library of Congress)

be assured that a privilege will be granted you for a period of fifteen years, even as M. Joly de Fleury has formerly indicated to you."[18]

Jouffroy was in despair. He had spent much of his money on his experiments. It seemed unfair that he should be called upon to make this terrific exertion. It was hardly possible that he could take his boat, large as it was, to Paris by way of the excellent system of French inland canals connecting the Saône, the Loire, and the Seine; moreover, the expense would be heavy.[19] The only way to meet the requirement was to transport the engine and other necessary equipment to Paris and there buy a boat in which they could be installed. Why the Academy of Sciences did not offer to bear expense in bringing the equipment to Paris, as it did earlier that same year in the case of the balloon, or why it did not send a committee to Lyons to inspect the boat is a mystery. The Royal Academy of Sciences at Paris, of course, was proud and stiff, and in this instance Périer, one of its members, has been suspected of jealousy. It is at least difficult to explain why Périer made no acknowledgment when Jouffroy sent him a miniature replica of his boat.[20]

Discouraged but still struggling, Jouffroy in vain sought aid from the Academy of Sciences. He even tried in the courts to force Follenay and others of his recently formed company to provide him with the funds necessary to carry out his purpose. This lawsuit was still unsettled when the Revolution came.[21] Friends urged him to take his invention to England, but he was too much the patriot for this. Ironically, during the Revolution he emigrated and fought under Bourbon standards, despite the mistreatment he had received. Not until the Consulate did he return to France. In the meantime John Fitch and James Rumsey in America and Desblancs and Robert Fulton in France had

[18] *Ibid.*, 166-67; Ballot, *Machinisme*, 394-95; *Nouvelle biographie générale*, XXVII, 45-48.

[19] See Joseph Dutens, *Histoire de la navigation intérieure de la France* (2 vols., Paris, 1829), I, 85-92; *ibid.*, II, iii. Dutens specifies the widths of each of the canals concerned. The locks between the side walls were only 4.4 meters (14.4 feet) in width.

[20] Figuier, *Merveilles*, I, 168. It does not appear that earlier in the century the Academy of Sciences made such a demand of the Marshal de Saxe in recognizing his two forms of treadwheel boats.

[21] Ballot, *Machinisme*, 395. According to Figuier, *Merveilles*, I, 168, Jouffroy continued to make boat trips on the Saône for sixteen months.

made other steamboat experiments, Desblancs at Trévoux and Fulton at Paris. Fulton admitted that he was inspired by the experiments of Jouffroy on the Saône in 1783. Bungling on the part of the government, or more particularly on the part of the academy, thus robbed France of much of the glory and profits of a major invention. As for Jouffroy, he died in 1832 an inmate of the Hôtel des Invalides, penniless and forgotten.[22]

During all this time the British had not been inactive. In 1736 Jonathan Hulls had obtained a patent for his design of a steamboat, and this plan was published the next year, but the boat was never constructed. In view of the imperfections of the steam engine of that day it is questionable that Hull's boat could have succeeded.

Later, in 1787-1788, an Edinburgh baker named Patrick Miller experimented with a paddle boat propelled by physical labor. On the suggestion of a friend he asked William Symington, an engineer who had just invented an improved steam engine, to apply his engine to the vessel. This was done in October, 1788, and the little boat, 25 feet long and 7 feet broad, equipped with two paddle wheels, plowed the waters of Loch Dalswinton at a speed of about five miles per hour. The next year Symington constructed another boat of approximately the same size, named the *Charlotte Dundas,* which under test pulled two heavy barges twenty miles in six hours against a strong wind. On the basis of these successful tests the British lay claim for Symington to the honor of inventing the first successful steamboat. The Duke of Bridgewater was so pleased with the experiments that he ordered eight vessels of the *Charlotte Dundas* type, but he died shortly afterward and his order was never filled. The *Charlotte Dundas* itself was left to rot at its mooring.[23] Thus experimenters of France, the United States, and Britain were working simultaneously at the invention of the steamboat. According to the time chart, Jouffroy's boat was the first to propel itself successfully.

In 1770, thirteen years before the demonstration of Jouffroy's boat, steam motive power had been applied to land transporta-

---

[22] *Nouvelle biographie générale,* XXVII, 48.

[23] Wolf, *History of Science,* 569-71; John Timbs, *Wonderful Inventions:* *From the Mariner's Compass to the Electric Telegraph Cable* (London, 1867), 251.

tion by Nicolas Joseph Cugnot (1725-1804), a military engineer born in Lorraine. Since his youth he had served in the army, first in Germany, later in Belgium, and finally in France. He had made some slight reputation by the invention of a new rifle, first used by French troops under Maurice de Saxe.

While serving in Belgium, Cugnot tinkered at constructing a steam-propelled carriage or truck, without success. The idea of an automatic carriage was by no means new. Several such carriages had been suggested earlier in the century and their plans approved by the Academy of Sciences. They did not, however, embody propulsion by steam. This feature, reportedly, had been suggested in 1680 by Isaac Newton and in 1759 by a Scottish student named Robertson, but prior to Cugnot no one had actually experimented with it.[24]

In 1763 Cugnot came to Paris as an officer in the French army and instructor in military matters. There he continued his experiments with a steam-propelled vehicle, using either the Brussels machine or a new one. In 1769 his experiments came to the attention of Gribeauval, inspector general of artillery. About the same time the Duc de Choiseul, minister of war and head of the French government, consulted Gribeauval about a similar machine designed by a Swiss officer named Planta. Planta was sent to inspect the machine of Cugnot, and told Gribeauval that it embodied the same features as his own. Choiseul then chose to support Cugnot's experimentation with government money. Previously he had been experimenting with a model, but Cugnot was now directed to work on a large scale, building a truck that could be used by the army in transporting artillery.[25]

On its completion this truck, carrying four persons, was demonstrated with success before Choiseul, Gribeauval, and a large number of spectators, moving at a speed of two and a half to three and a half miles per hour. Choiseul was pleased with the performance and ordered Cugnot to construct a second and larger truck capable of making 1,800 *toises* (2.17 miles) an hour and of transporting a load of four or five tons. While working on this

[24] *Nouvelle biographie générale*, XII, 591; *Machines approuvées par l'Académie*, III, 33-40; *ibid.*, V, 171-73; *ibid.*, VI, 141-44; Hodgins, *Behemoth*, 136; Figuier, *Merveilles*, I, 263.

[25] Ballot, *Machinisme*, 388; Figuier, *Merveilles*, I, 263-64.

second truck Cugnot made further tests with the first model, and it is related that in November, 1770, it carried a load of two and a half tons (a cannon base) and traveled at the rate of three and an eighth miles an hour. Rugged terrain offered no impediment to it.[26]

The second, larger truck was completed in July, 1771, at a cost to the government of 20,000 livres. Unfortunately Choiseul had fallen from power in 1770, and his successor showed no interest in the new invention. There is no certain proof that it was ever demonstrated. It was placed under a shed at the Arsenal, where it long remained.[27]

In 1798, between his Italian and Egyptian campaigns, Napoleon showed an interest in the machine and called it to the attention of the Institut de France; but shortly afterward he set out for Egypt, and the Institut forgot the truck. The next year, however, Molard, director of the Conservatoire des Arts-et-Métiers (Museum of the Arts and Trades), "requested the steam truck *(chariot à vapeur)* for this establishment." Due to opposition from the minister Roland and certain officers, the request was not granted until 1801, when the second and larger of the two trucks was placed in this scientific museum, where it remains to this day.[28]

The truck has three wheels—a small one in front and two large ones at the rear. All are thick and sturdily built, but the front wheel is too weak for the functions it had to fulfill. For not only was it to bear an undue portion of the load but also to be used for steering. Its tricyclical construction made the machine easily overturned. The boiler and cylinders, made of copper, protruded over the front wheel and gave the appearance, as one book has described it,[29] of a whisky still on a wheelbarrow. Only enough water could be carried in the boiler to last about fifteen minutes, and there was no way to put more in it without stopping,

---

[26] Ballot, *Machinisme*, 388; Figuier, *Merveilles*, I, 264. According to a tradition that cannot be verified, the truck ran into a section of the wall surrounding the Arsenal, breaking it down.
[27] Ballot, *Machinisme*, 388.
[28] *Ibid.*, 389; Figuier, *Merveilles*, I, 264. The present writer saw the machine in the Building of Scientific Exhibits at the Paris Exposition of 1937.
[29] Hodgins, *Behemoth*, 137.

for the truck had no tank for extra water. Even had there been such, there was no way of adding water gradually and continuously while the truck was in motion. Beneath the boiler was a fire pan, but its embers would not last long and there was no fuel bin for extra coal. Here again one would have to dismount from the truck to add fuel. Clearly enough the machine had weaknesses that ought to have been foreseen. The boiler and fire pan should have been so placed that they could be refilled by the crew without dismounting and without halting the truck, and the truck should have been equipped with water tank and coalbin. In all likelihood these would have been added, if experimentation had continued. Yet these added features would perhaps have left little room on the truck for the field gun or its support, and a trailer would have had to be devised.

As it was, the machine was built with more cleverness than these criticisms might imply. The steam from the boiler passed into two cylinders[30] whose pistons were connected with the front wheel by means of ratchets and levers. The pistons worked alternately, giving a continuous movement. Around the front wheel was a band of iron, fluted to give it greater drawing power. Beneath the fire pan Cugnot had placed a layer of loose earth as a sort of insulating pad to hold the heat of the embers.[31]

The machine had its merits and its weaknesses. There were some who praised it and some who scoffed, depending upon their ability to foresee its potentialities. Here again the French had within their grasp one of the most useful inventions of modern times—the steam-propelled vehicle—but a freakish turn in politics threw its promoters into the discard and left its inventor without support.

Cugnot indeed was given a pension of 600 livres a year, paid until the Revolution.[32] Then he was left without income and would have perished had not a generous woman in Brussels, where he emigrated, provided for him. Later under the Consulate, L. S. Mercier, author of *Tableau de Paris*, called the atten-

[30] Boutaric, *Inventions françaises*, 222.
[31] Figuier, *Merveilles*, I, 265-66.
[32] Granted in 1779, on the funds of the ministry of war. *Archives parlementaires*, XIV, 85.

tion of Napoleon to his condition, and Cugnot was given a pension of 1,000 livres until his death.[33] He was the author of three treatises on military engineering, but it is the steam truck that has immortalized his name.

Even before Cugnot's death (1804) the British were performing wonders with his invention. William Murdock, Scottish assistant to James Watt, was inspired to pursue Cugnot's idea further and constructed a model steam cart (1784) fifteen inches high, which ran successfully on British roads at six to eight miles per hour. Richard Trevithick in turn was inspired by the success of Murdock, and in 1801 and afterward he built several full-sized locomotives. A few years later, Gurney attempted to run a bus service with his steam coaches, and George Stephenson perfected the locomotive and took over the idea of a railroad. After 1800, developments moved swiftly in transportation by steam.[34]

[33] *Nouvelle biographie générale*, XII, 491; *Biographie universelle*, X, 335.

[34] Wolf, *History of Science*, 554-55.

CHAPTER III

The Telegraph

THE FIRST ELABORATE TELEGRAPHIC SYSTEM WAS ALSO THE invention of eighteenth-century France. Throughout history, communication at a distance has been made by fires, lights, pigeons, and other means. Prior to the invention of the telegraph, various suggestions had been offered and attempts made to arrive at a complex method of distant communication through sight or sound. The Jesuit Flamianus Strada, in the early 1600's, greatly impressed with the work of William Gilbert in magnetism, suggested that two men at a distance might communicate with one another by the use of magnetic needles pointing toward letters on a dial. Joseph Addison in the *Spectator* revived in 1712 this idea of the "sympathetic" needles. The 1700's witnessed suggestions and experiments by several men of many nations who would use electrical means. This group included Charles Morrison (1753), a Scottish surgeon; George Louis Lesage (1774), a Swiss physicist at Geneva; Lhomond (1787), a French physicist; Reisser (1794), a German; Betancourt (1781), a French engineer in Spain; and Francisco Salva (1796), a physician in Barcelona. Of these, Lesage reportedly obtained the most encouraging results, but neither his instruments nor those of the other experimenters in electrical telegraphy at that time were really more than objects of cabinet interest.[1] Despite its popular

[1] See accounts of these men in Figuier, *Merveilles*, II, 88-92; Wolf, *History of Science*, 662-64; *Dictionary of National Biography*, ed. by Leslie Stephen and Sidney Lee (22 vols., London and New York, 1908-1909), XIII, 1004 (hereafter, *D.N.B.*); Arthur Young, *Travels in France by Arthur Young during the Years 1787, 1788, 1789*, ed. by M. B. Betham-Edwards (2d ed., London, 1889), 96. Young gives this description of the apparatus and experiments of Lhomond, which he witnessed on a visit to Lhomond's home in Paris October 16, 1787: "In electricity he has made a remarkable discovery: you write two or three words on a paper; he takes it with him into a room, and turns a machine inclosed in a cylindrical case, at the top of which is an electrometer, a small fine pith ball; a wire connects with a similar cylinder and electrometer in a

interest and rapid strides in the eighteenth century, electricity was still too little understood to become the means of telegraphic communication.

While the world in this respect awaited appropriate developments in electricity, a French abbé, Claude Chappe (1763-1805), with the aid of his three brothers, invented in the early 1790's a mode of visual telegraphy at once simple yet elaborate, capable of transmitting messages of any nature over a fixed network of stations, day or night.[2]

All four brothers were given college education by their father, a man in comfortable circumstances financially. Claude, destined to take the lead in the invention of the telegraph, attended first the Collège de Joyeuse at Rouen and later the seminary at La Flèche. Entering the priesthood, he obtained a well-paying benefice which enabled him to live at Paris and to devote his time to research in physics, above all in electricity where his interests particularly lay. Losing this benefice early in the Revolution, he returned to his family at Brûlon, near Le Mans and La Flèche. There he turned his attention to the problem of telegraphy. For a time he worked at an electrical solution, but without success. Later in 1790 he and his brothers experimented at sending messages between two stations a quarter mile apart by means of two posts to each of which a pendulum was attached. On March 2-3, 1791, they gave public exhibitions in the department of Sarthe, sending messages between two stations 15 kilometers apart.

---

distant apartment; and his wife, by remarking the corresponding motions of the ball, writes down the words they indicate: from which it appears that he has formed an alphabet of motions. As the length of the wire makes no difference in the effect, a correspondence might be carried on at a distance: within and without a besieged town, for instance: or for a purpose much more worthy, and a thousand times more harmless, between two lovers prohibited or prevented from any better connection. Whatever the use may be, the invention is beautiful."

[2] The story of a visual system of telegraphy in the reign of Louis XIV, invented by the well-known French physicist Guillaume Amontons (1663-1705) and tried with success over short distances, is related by Walter Lodian, "A Century of the Telegraph in France," in *Popular Science Monthly*, XLIV (1893-1894), 791-92. According to Lodian, one experiment was conducted in the presence of the dauphin and another before the Duchess of Orleans. Despite the success achieved, the government refused to support the enterprise, regarding it merely as a means of amusement, and Amontons abandoned work on it.

Later in 1791 the brothers went to Paris for a demonstration. They set up a station at the Place de l'Etoile on the Champs Elysées, but at night it was torn down by hostile parties. Next they erected a station in a private park at Menilmontant, but shortly it, too, was destroyed, the revolutionary populace thinking it a device by which a royalist might communicate with the king and his friends. By this time Chappe had already spent 40,000 livres of his own money in experimentation. He now sought police protection for his towers from the Legislative Assembly, in which one of his brothers, Ignace, was a deputy.

He also requested permission to appear before the Assembly to present his invention, granted for May 22, 1792. At the hearing he told how speedily messages could be sent, day and night. The Assembly recognized his invention and asked the committee of public instruction to examine its merits, but nothing came of it inasmuch as the Assembly came to an abrupt end in September.

Later, Chappe won the attention of the National Convention. On April 1, 1793, his invention was favorably reported by Romme, speaking for the joint committees of war and public instruction, as deserving a test. If it measured up to its possibilities as seen by the committee, it might be a useful war aid in facilitating the rapid exchange of dispatches between the government and its armies. Several other suggestions of quick communication had been proposed to the committees by one person or another, Romme stated, but this seemed the only one deserving of consideration. He added that Chappe's invention had already been tried between two communities in the department of Sarthe with satisfactory results. Chappe had appeared for a conference with the committees and had satisfactorily answered all questions, save one. Could it be used in a fog? This was its single weakness. Anticipating the query whether the nature of messages could be kept secret, Romme explained that a code was used for words, so that none but station agents familiar with it could understand what was being sent. This permitted the transmission of messages of a confidential nature. In conformity with his request, the Convention ordered an inquiry into the new invention, asking its committee of public instruction to appoint commissioners for this purpose, and set aside 6,000 livres from war funds for this experiment.³

To this committee were appointed Lakanal, Arbogast, and Daunou, the first two being scientists. They made a test on July 12, as later reported by Lakanal, messages being sent between two stations eight and a half leagues (approximately 21 miles) apart. There was some difference of opinion among the committee members on its merits, with Lakanal strongly impressed with its possibilities. On July 26 he presented a highly favorable report to the Convention. On the basis of messages sent between the two stations on July 12, he predicted that a dispatch of ordinary length might be sent from Valenciennes, in French Flanders, to Paris in thirteen minutes and forty seconds— a remark which brought applause from the Convention. The erection of a chain of sixteen stations to connect Paris with the Army of the North at Lille would cost about 58,400 livres. The committee favored this project and taking over the invention for the benefit of the nation while reimbursing Chappe. It recommended that Citizen Chappe be given the title of *ingénieur-télégraphe* (telegraphic engineer) and the pay of a lieutenant in engineering.⁴

The Convention adopted the proposed measure and ordered construction of a telegraphic line of towers from Paris to Lille. Chappe himself was placed in charge of the undertaking, and allowed priorities on metals, timber, and stone. He set about accumulating stocks of each, but is said to have been careless in business methods, keeping no records of expenditures and making payments through third parties. Nevertheless the work moved smoothly, his finances were never questioned, and the line was completed in August, 1794. No further sabotage from the populace occurred, inasmuch as the government had given the undertaking police and military protection since July 2, 1793.

Many messages were sent from station to station while the work was under construction. The first recorded message to be sent after the line was completed to Lille was August 15, 1794, announcing to the committee of public safety the capture of Quesnoy from the Austrians. The news was known in Paris one

³ *Moniteur universel*, XVI, 30-31.   ⁴ *Ibid.*, XVII, 250-51.

hour after the troops entered the town.⁵ On August 28, the report came over the new line that Valenciennes had been taken. But as yet it produced no great public excitement. Success for so many things, in that day as in this, depended upon dramatization.

For telegraphy, this came on September 1, 1794, in connection with the recapture of the town of Condé from the Austrians. L. N. M. Carnot, minister of war, mounting the tribune in the Convention said: "Citizens, here is the news just received this instant by the telegraph that you have established from Paris to Lille: 'Condé is restored to the Republic; the retaking took place this morning at six o'clock.'" Such was the terse message telling of another defeat to the large Austrian army of 80,000 men in French Flanders. The reporting was dramatic, and the Convention went wild with applause. One deputy requested that the name of the town be changed to that of Nord-Libre (evidently because "Condé" was too Bourbon for the republicans). This was decreed. Another deputy proposed that the victorious army be congratulated. Accordingly the Convention sent a return message, saying that the army deserved well of the country and that Condé had a new name. The message was telegraphed to Lille and from there sent by courier to Condé. Before the Convention had concluded that morning's session, word came back from Lille that the message had been delivered.⁶

The Convention now was enthusiastic about the new invention. It ordered an extension of the line from Lille to Ostend and the construction of a second line to Landau in Bavaria via Metz, but shortage of coin for payment of workers and of stocks of metals and wood caused construction to bog down. The Directory which came into power in October, 1795, showed the same zeal for telegraphy as had the Convention, and the work continued. The second line was completed to Strasbourg by 1798. It consisted of 46 towers (as against the 16 in the line from Paris to

---

⁵ Edme Champion, "Le premier télégramme," in *Révolution française*, XLII (1902), 53-54; P. Mantouchet, "A propos du télégraphe Chappe," *ibid.*, LXXXVII (1934), 260-62. Quesnoy was a few miles north of Lille, to which the message was brought by courier.

⁶ Figuier, *Merveilles*, II, 39. Figuier and others describe this as the first telegram sent over the completed line; in this, however, they are mistaken.

Lille) and cost a total of 176,000 francs.[7] In 1798 also, the line to Lille was extended to Dunkirk; a third line, from Paris to Brest, was undertaken;[8] and a fourth line, from Paris to Lyons, via Dijon, was ordered. Work on this last line went slowly, however, and it was not in operation by 1800.[9]

The lines did not bring in monetary returns, but on the contrary absorbed great outlays. Maintenance and service alone in the year VIII (September 23, 1799–September 23, 1800) cost 434,000 francs. Faced with this situation, Napoleon, whom one might readily have expected to favor the new invention because of its military potentiality, showed himself lukewarm and in December, 1800, reduced its appropriations to 150,000 francs a year.[10] Because of this cut, the line from Paris to Lyons had to be abandoned.[11] Chappe was in despair lest his invention, so full of promise, should now collapse. A strategem occurred to him. It might be used for business even as for war—for sending stock quotations, news of the arrival of ships, and reports of the lottery. France had been plagued with an abuse in connection with the lottery. Men of little scruple would set out from Paris after the lottery drawings and sell or buy tickets in the provincial towns before the reports of the drawings arrived. To the managers of the lottery, Chappe pointed out that the telegraphic lines would help to get rid of this nuisance. They seized the idea and placed considerable sums at Chappe's disposal.

During the Restoration, interest in the telegraph greatly increased, and by the middle of the century, when the system was still in use, Chappe's dream of seeing Paris connected by chains of stations to all its frontiers and leading seaports had been realized. Sadly enough, Chappe committed suicide in 1805 while

---

[7] The stations of the Paris-Lille route averaged 14 kilometers apart, a distance too great for the best results. In later lines, the stations were placed somewhat closer. The telegraphic station at Strasbourg was in the cathedral bell tower. *Moniteur universel*, XXIX, 117.

[8] It was completed in seven months at a cost of 300,000 francs. There were 55 towers. Figuier, *Merveilles*, II, 42.

[9] For the construction of these lines, see also Bast, *Merveilles du génie de l'homme*, 156-57; A. Rambaud, "L'agriculture, l'industrie, le commerce, le crédit pendant la Révolution," in *Révolution française*, XIII (1887), 234-35.

[10] It must be observed, however, that money had a higher value under Napoleon than under the Directory.

[11] Construction of lines was expensive. The line to Strasbourg, completed in 1797 or 1798, with 46 towers, cost 176,000 francs. Figuier, *Merveilles*, II, 42.

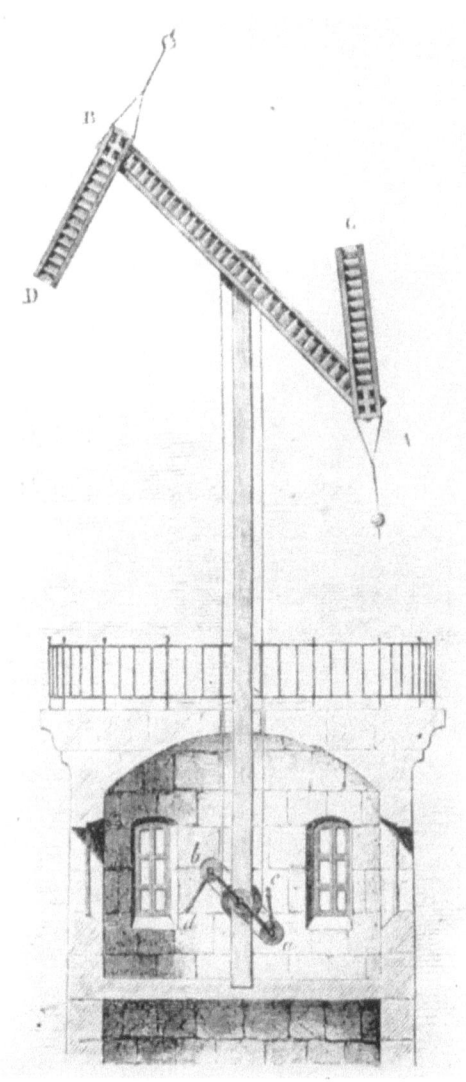

THE INTERIOR OF A SEMAPHORE TELEGRAPH TOWER

The mechanism was invented by the Abbé Chappe in 1794.

(Illustration from Figuier, *Merveilles*, II, 52, courtesy of the Library of Congress)

in a fit of despondency caused partly by the slow progress of his invention and partly by painful bladder trouble.[12] After his death two of his brothers, Ignace and Pierre, assumed control of the telegraphic lines, while the third, Abraham, worked at a means of overcoming fog. Of interest is the fact that Abraham anticipated to some degree methods of English airfields in the war of 1939-1945, using hydrogen fires to burn away the fog.[13]

The stone towers employed in Chappe's system resembled the windmills of that day. They enclosed a small room where the operator sat (and in some instances slept), protected from the weather. Rising from the tower to a height of ten feet or more was a heavy pole, near whose top was a strong crossbar, approximately fourteen feet in length, working on a pivot. To the ends of this bar, called a regulator, were attached two arms, about six feet long, which could be turned at any angle desired by means of wires extending down into the operator's room to a control board representing in miniature the positions of the regulator and arms overhead.[14]

The regulator and arms could be placed in 196 positions other than those where the regulator would be vertical or horizontal. Chappe decided that no signals would be sent save when the regulator was moved to an inclined position, serving to notify the next station that a message was coming. Of the 196 positions, he chose the 98 made when the regulator was inclined to the left *(oblique de gauche)* for use only between operators, the other 98 *(oblique de droit)* would be used for dispatches. With the aid of a friend, Léon Delaunay, who had long been in the French consular service at Lisbon and had studied diplomatic cipher, Chappe formulated a code of 9,999 words, later reduced to 8,464, each represented by a number. These were set forth in a code book of 92 pages, each page carrying 92 numbers and words. When one station agent wished to communicate with another, he called the number of the page with his first signal and the number of the word on the page with the second. The operators therefore had to be conversant with the code, and speed of transmission depended upon their deftness in using it. The Chappes

[12] *Ibid.*, 43-44.
[13] *Ibid.*, 45.
[14] *Ibid.*, 51-52; Wolf, *History of Science*, 661.

likewise formulated a phrase vocabulary and a place-name vocabulary, each consisting also of 92 pages of 92 items apiece. When using these second and third codes, the transmitting operator by his first signal called the code, by his second the page, and by his third the item on the page. Other words, it appears, were spelled, longer time being required.[15]

For night messages four lanterns were placed at the ends of the signaling arms. Thus by night as well as by day, and in all types of weather save fog, messages could be transmitted by this system.

The very language of the telegrams *(bulletins télégraphiques)* has a modern ring. Here is the report of one recorded in the *Moniteur*, the official government journal, as read to the Convention 5 brumaire an III (October 26, 1794): "Bulletin télégraphique of the 4th Brumaire, from 4:30 to 5:20; transmission from Lille, composed on the vocabulary of the engineer:

"'Hulst, Sas-de-Gand, Philippine, and Axel [four places in Belgium] have been in the possession of the Republic since the 2nd of this month. A single Frenchman has been dangerously wounded.' Signed, Chappe, telegraphic engineer."[16]

By a strange coincidence, a Swede named Endelcrantz worked out a similar visual system of telegraphy in 1794, the first trials being made over a line from Stockholm to Drottingholm on October 30. The English, too, worked out a system using shutters in 1796, and a line was constructed from London to Dover by the Admiralty; but despite improvements it was never satisfactory.[17] All three systems continued in use until superseded by electrical telegraphy in the nineteenth century.

The telegraph system of Chappe lasted well into the nineteenth century, and was highly developed in Western Europe when the electrical telegraph made its appearance. Only by degrees was it replaced by the latter. Nor was it so inferior as one might

---

[15] Figuier, *Merveilles*, II, 23, 52-53. These three vocabularies were used until 1830, when they were united into one.

[16] *Moniteur universel*, XXII, 356.

[17] Figuier, *Merveilles*, II, 58-59, gives an account of these systems. A second British system was created by Richard Lovell Edgeworth, who by 1767 reportedly had been sending messages to friends over short distances. Not until 1797 or 1798, however, was a line of this type constructed, in Ireland in anticipation of enemy invasion. When danger of invasion passed, the line was neglected. *D.N.B.*, VI, 383-84.

hastily conclude. The *Popular Science Monthly* in 1894, on the centenary anniversary of Chappe's invention, carried an article which stated that more time was required even then for sending a message by electric telegraph than by Chappe's aerial telegraph.[18]

As for the speed acquired by the aerial telegraphers in the 1860's, when he was writing, Figuier says that only three minutes were required for transmission of messages from Calais to Paris, a distance of 68 leagues (170 miles), via 33 stations; two minutes for messages from Lille, 60 leagues (150 miles) distant, via 22 stations; eight minutes for messages from Brest, 150 leagues (375 miles) distant, via 54 stations; and twenty minutes for messages from Toulon, 267 leagues (667 miles) distant, via 100 stations.[19]

[18] Lodian, "Telegraph in France," 795.

[19] Figuier, *Merveilles*, II, 54.

Chapter IV

Lighting

EIGHTEENTH-CENTURY FRANCE GAVE BIRTH TO A NOTABLE series of inventions in lighting. So excellent are the lighting facilities of our own day that it is difficult to believe that they are of recent origin. At the beginning of the eighteenth century Western Europe used only a crude lamp of ancient origin, the candle, which had come into existence in the Middle Ages, and the lantern, also dating from the Middle Ages. Their light was poor. The lamps did not have the glass chimneys of later days; while the lanterns, though enclosed with glass, had defects in feeding both with oil and air, and wicks were inclined to burn unevenly and the glass to fog with soot. There was a slight amount of street illumination. The lighting of Paris had begun around 1667 under Lieutenant of Police La Reynie, and only in 1697 did Louis XIV order that other leading French cities follow Paris's example. The lanterns used for the purpose were about twenty inches high and twelve inches square. They enclosed a large candle and a metal reflector, and were suspended by a cord from a beam some nine or ten feet high. These lanterns, about twenty feet apart, were lighted regularly, not only on dark nights but also when the moon was bright. Travelers praised them as preferable to the lanterns of London, which were burned only fifteen days in the month.[1]

Other forms of lantern for street lighting had been devised during the first half of the eighteenth century, but not adopted. In 1703 the Royal Academy of Sciences had approved a lantern invented by Favre, in which a set of four lamps facing the cardinal points and fed by a single oil container would supposedly give a more abundant light. Each lamp sat in a parabolic metal basin designed to throw the light a greater distance than the

---

[1] Figuier, *Merveilles*, IV, 2-11. In 1729 Paris streets were provided with 5,772 lanterns. Bast, *Merveilles du génie de l'homme*, 167.

lanterns in use.² In 1744 a reflecting oil lantern invented by Bourgeois de Châteaublanc was approved by the same academy. This lamp was cone-shaped, the cone consisting of glass and pointing downward. Above it was a metal cover of tin or copper, and at its topmost part a ring by which it could be suspended by a chain or cord. Within the glass cone was a smaller metal cone holding the oil and wick, fed by two metal tubes from above. At the top of these tubes was an opening into which an attendant could pour oil without removing the glass. Moreover, there were small air holes for feeding the lamp with oxygen. This lamp seems to have been of practical nature, but apparently it was not used.³

In 1764 the Academy of Sciences offered a prize of 2,000 livres for the best paper to be submitted the next year on "The Best Means of Lighting the Streets of a Large City During the Night, Combining Clearness, Facility of Service, and Economy." Papers with new designs of lamps were submitted by several scientists later to be renowned—Lavoisier, Bailly, Leroy, and Bourgeois de Châteaublanc. The committee of selection was wearied by Lavoisier's paper, which they regarded as too theoretical, and divided the prize among the other three contestants. Nevertheless, the academy published Lavoisier's paper and the king was asked to award Lavoisier a gold medal. The lantern preferred was an invention by Bourgeois de Châteaublanc (1698-1781), fed by oil and carrying a metal reflector. Actually it was little more than a modification of his lamp of 1744 and the reflecting candle lamp already in use. It was adopted for lighting Paris, and Bourgeois de Châteaublanc was given a small pension. Though he was an applicant for control of the lighting of Paris in 1769, another was preferred over him and given a twenty-year lease.⁴ Bourgeois' pension was not even paid with regularity, and he had trouble over his inventor's right.⁵

The next development was more important—the oil lamp of Aimé Argand. This was a circular burner and wick which could

---

² See illustration and description in *Machines approuvées par l'Académie*, II, 53-54.
³ See illustration and description *ibid.*, VII, 273-74.
⁴ *Nouvelle biographie générale*, VII, 79, says that he obtained the twenty-year lease for lighting Paris in 1769 but that he was defrauded out of it by two assistants.
⁵ Figuier, *Merveilles*, IV, 12-13. An illustration of this form of lantern,

be raised and lowered at will, together with a cylindrical glass chimney.[6]

The inventor, Aimé Argand (1755-1803) was Swiss-born and educated. His father, a Genevan watchmaker in comfortable circumstances, gave his son a liberal education at the Academy or University of Geneva. One of the boy's instructors was the celebrated physicist Horace Bénédict de Saussure. Though the parents wished their son to enter the clergy, he preferred physics and chemistry. With this training he set out in 1775 for Paris to make his way in the world. It is clear that he went with high recommendations, for in the next year he was asked to read a paper before the Academy of Sciences, no mean honor for a youth. He became closely associated with Lavoisier and Fourcroy, to whom De Saussure had recommended him, and through their assistance and guidance gave a course of lectures on chemistry, more especially on distillation. Among his auditors were some winegrowers of Languedoc who became greatly interested in his criticism of the methods of distillation in vogue in their province. They were struck with the better method that he proposed and prevailed on him to accompany them to Montpellier and demonstrate his superior method before De Joubert, treasurer of the province. This he did in 1780, and perhaps by way of caution invited one of his brothers to meet him there. The demonstration went well; De Joubert was impressed; the Royal Academy of Sciences of Montpellier received a favorable report from its committee of judges on the "new distilleries" of Argand; and a committee of judges outside the academy, headed by the Abbé Rozier, sent a favorable report to D'Ormesson, controller general of finances at Paris. The royal government directed the intendant at Montpellier to make a thorough examination of the new procedure. This was done, and the Argand brothers emerged with royal decorations and 120,000 livres from the province in purchase of their rights as inventors.[7]

---

wrongly called "le premier réverbère," is shown on p. 14. It was hexagonal, shaped like a bucket with a domelike metal cover and a handle with a small pulley at the top, through which a wire was passed to hang it.

[6] See description *ibid.*, 22-23.
[7] *Ibid.*, 15-16.

While at work in Lower Languedoc on this invention, Argand conceived of a new form of wick for oil lamps which would add greatly to the light.[8] In 1780 he made his first model and placed it in one of the distilleries on the property of De Joubert. Saint-Priest, the intendant of Languedoc, took an interest in the lamp and presented it to the Estates of Languedoc, whose members admired it, even though at this stage it had a metal rather than a glass chimney. Shortly afterward (January, 1784) in Paris, Argand invented a glass chimney and thereby completed his lamp.[9]

A friend of the Montgolfiers, Argand visited them in Annonay in 1783 and accompanied Joseph to Paris preparatory to his exhibition before the court. Each day he met with him in the gardens of Reveillon, with whom Montgolfier was staying, in the Faubourg Saint-Antoine. Among the scientists and others of note whom he met there were two members of the Academy of Sciences, Cadet de Vaux and Lesage, to whom he described his new lamp. They were greatly impressed and insisted that he display it to Lenoir, lieutenant of police, so that it might be adopted to light the Paris streets. Together the three scientists called on Lenoir, who received them and was so pleased with the new lamp that he wished at once to use it. He asked about its construction, but Argand, not protected by patent, refused to divulge information on this matter. Thereupon Lenoir rashly refused to discuss the matter further.

---

[8] *Ibid.*, 16.

[9] *Ibid.*, 18, 21, 28. Various writers have suggested that Argand might have borrowed ideas from others. Henry René d'Allemagne, *Histoire du luminaire depuis l'époque romaine jusqu'au XIXᵉ siècle* (Paris, 1891), 370, points out that lamps using the "double current of air" had been made by Stamius in 1750 and by Jacques Jégot, a canon of Troyes, in 1774, and that some *réchauds* for heating spirits of wine then in use operated on the same principle. He believes that Argand borrowed the idea from these last, but was original in using a circular rather than a straight wick. Figuier mentions the possibility of Argand's indebtedness to Captain (later General) J. B. M. Meusnier (1754-1793), who thought of such a lamp about the same time as did Argand. Proof of indebtedness to others, however, is lacking, and Argand is still credited with this invention.

Lamp chimneys had been in existence at least since 1717, when the government granted Jean Le Vaillant a monopoly in blowing them. Warren C. Scoville, "State Policy and the French Glass Industry, 1640-1789," in *Quarterly Journal of Economics*, LVI (1941-1942), 436.

Why Argand did not at this time apply to the Academy of Sciences for approval of his invention and thus obtain protection for fifteen years is not indicated. It was still apparently before Meusnier had presented the report on his own lamp to the academy. Whatever the answer to this mystery, the fact remains that Argand did not apply. Instead, perhaps in pique, he went to England and there patented his lamp in 1783. Later he returned to Paris.[10]

Now entered two imposters on the scene, Quinquet and Lange. These two men had come to know Argand in the gardens of Reveillon where Montgolfier was at work on his balloon, heard of his remarkable lamp, and determined to get its secret from this new Aladdin. With bloodhound tenacity they probed him and everyone who knew him about details of his lamp, and they succeeded in learning enough to duplicate it. Many persons by this time, of course, had seen the lamp in operation—De Joubert, members of the Estates of Languedoc, Fautras de Saint-Fond, Cadet de Vaux, Lesage, Lenoir, officials in England, and perhaps others. De Joubert, in fact, in 1783 had ordered a tinsmith in Paris to make him such a lamp from designs that he sent. Count de Milly of the Academy of Sciences learned what the two men were about, and to protect Argand cited him in a memoir to the Academy on January 21, 1784, as "the author of this ingenious lamp," for whom he wished to conserve all rights. Some time after this the imposters presented their copy of the lamp to Lenoir, who at once recognized it and stated, "This is the lamp of M. Argand." The two men did make an improvement in the chimney. Where Argand's chimney had straight walls, that of Quinquet and Lange was compressed above the wick, so that the upper portion of the chimney was smaller than the lower—a feature that was commonly followed in the future.[11]

News of the poaching on his invention led Argand to break short his stay in England and rush home to protect his rights. While his lamp wick (the major feature of his invention) was not protected by government recognition in France, his glass chimney was. This had been recognized by an order of the Council of State of August 30, 1785, registered with the parlement

[10] Figuier, *Merveilles*, IV, 19.   [11] *Ibid.*, 22-24.

of Burgundy. By it he was given the exclusive right to manufacture and sell glass chimney for lamps under the name of *lampes d'Argand* for a period of fifteen years. Argand brought suit in the courts, but judicial procedure was slow, and in time he considered it advisable to compromise with his adversaries and to enter into a business alliance with them. He allowed them to establish themselves at Paris, while he went to Versoix, near Geneva, and undertook to manufacture lamps for southeastern France and Switzerland.

Letters patent dated January 5, 1787, gave Argand and Lange the "exclusive permission to manufacture and sell, throughout the kingdom, the lamps of their invention for fifteen years."[12] No sooner was this legal process settled, however, than the tinsmiths of Paris brought suit against them, claiming that the privilege violated their rights. The tinsmiths were defeated, but hardly was their case ended when the Revolution came and all privileges allowed industrialists were revoked, and "the manufacture of lamps with double current of air and a glass chimney, fell into the public domain."[13]

Argand left first for England, and later for Switzerland. His marriage proved unhappy, his only child died in 1794, in melancholy he turned to occultism, and in poverty he died in Geneva in 1803. Lange and Quinquet remained in Paris and made a success manufacturing the lamp which came to be called, paradoxically, the "Quinquet lamp." The term "Argand lamp" was given a lamp with a metal chimney. Such was the fate of an inventor![14]

Argand's wick and glass chimney revolutionized the procedure of lighting. The lamp was a great improvement over anything in use in its day, giving a much brighter flame than other lamps because its circular wick, as Argand explained, received a double current of air. Even so, the lamp still needed improvement. While free from smoking so long as its wick was kept trimmed, the wick did have a faint tendency to carbonize, as feeding was

[12] *Ibid.*, 25.
[13] *Ibid.*, 26. Figuier gives the date of the tinsmiths' suit as 1789; Paul M. Bondois, "L'industrie et le commerce sous l'ancien régime," in *Revue d'histoire économique et sociale*, XXI (1933), 180, says that it was 1787.
Argand was also sued in London by the glassmakers in 1786. Figuier, *Merveilles*, IV, 26.
[14] Figuier, *Merveilles*, IV, 26.

not as rapid as it might have been. In common with other French lamps of the time it had an oil tank or reservoir attached by a horizontal tube to feed the wick. This tank naturally caused a shadow when the lamp was lighted. To prevent it, a Frenchman named Phillips late in the century devised a "circular reservoir around the wick, placing this reservoir on the same level as the wick." "The wick was thus placed in the center of the oil reservoir." This lamp Phillips called the *Lampe Sinombre*—the Shadowless Lamp. Later lamps employed this feature.[15]

These developments paved the way for the next invention—the most notable lamp of the eighteenth century—the Carcel lamp of 1800. This lamp was provided with a mechanism which pumped oil from the reservoir below the wick into the basin where the wick rested so that there was always about the wick a tiny lake of oil. The result was an intense light and a complete absence of wick carbonization.

This masterpiece of oil illumination was the achievement of an obscure Paris watchmaker, Guillaume Carcel (1750-1812), who longed to be an inventor and after years of experimentation found the key that he sought, a small quadrilateral box provided with a piston and pump that he placed just above the oil reservoir in the base of the lamp. This apparatus sent through a tube a constant, even flow of oil to the wick above. The pump box was deftly arranged. It consisted of three floors or levels connected by canals and floodgates. On the lower floor were two valves, one at each end, that opened themselves on suction from above to draw more oil from the reservoir. In the roof of this lower floor were two open canals, one at each end, to permit the oil to rise under force of suction to the middle level. In this middle level a piston, attached to a clock outside the lamp and timed by it, moved horizontally. As the piston moved in one direction, pushing the oil in the middle chamber and forcing it through a valve to the third or top level, suction was created behind it on the middle level, thereby drawing oil to the middle level from the one below. The piston, of course, had a double movement, and as it returned it forced the newly sucked oil of the middle level to the third level through a second valve. Thus each of the

[15] *Ibid.*, 31.

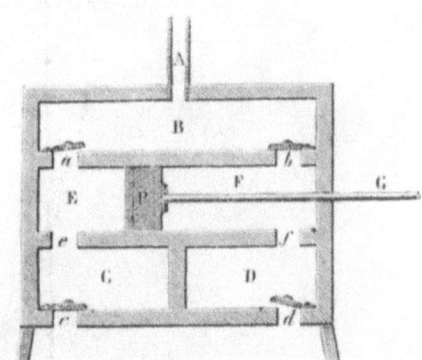

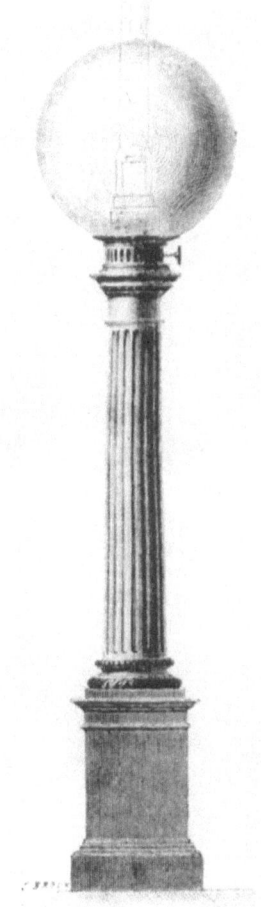

THE CARCEL LAMP

This most brilliant lamp of the eighteenth century was patented in 1800. The diagram shows the mechanism of the fuel pump.

(Illustrations from Figuier, *Merveilles*, IV, 36-37, courtesy of the Library of Congress)

three floors had two openings through which oil could be raised. The first and third floors were equipped with valves, so that the oil could be forced only upwards. From the middle of the third level a tube ran to the basin of the wick. Thus when the lamp was lighted, the piston was constantly sending a regular flow of oil to feed the flame.[16]

A neighbor and friend, the pharmacist Carreau, with whom Carcel shared the joy of his new invention, offered to furnish the money for marketing the new lamp, and Carcel applied for a patent *(brevet d'invention)* in their joint names. This was given October 24, 1800.

The new lamp, instead of being called by some term like "Glow-worm," "Firefly," or "Midnight Sun" which might have helped in gaining it popular recognition, was afflicted with the Greek derivative *Lycnomènes* ("Steady Light"). For this Carcel was responsible. The lamp did not sell. Perhaps it was due to the fact that France was too immersed in war—for only in a short interval (1802-1803) was she not at war with some country until Napoleon's fall. Carreau saw his money going down a rathole and withdrew from the partnership at an early stage. Carcel carried on alone. There was one fleeting moment of glory for him when early under the Empire, Napoleon opened an Industrial Exposition on the Champs-Elysées and Carcel was asked to exhibit his lamps. This he did with great pride and received the plaudits of the populace, but apparently made few sales. Of course it was a luxury lamp, priced out of reach of the masses. After Napoleon was removed in 1815 a swarm of nobles came back to France—a France at peace—and these men avidly bought Carcel lamps for their mansions from one or another of a half dozen manufacturers. But Carcel had died in 1812 in poor circumstances, failing to enjoy the glory and opulence which his

[16] *Ibid.*, 32-36; Charles Ballot, "Philippe de Girard et l'invention de la filature mécanique du lin," in *Revue d'histoire économique et sociale*, VII (1914-1919), 146-47; Jean Baptiste Joseph Delambre, *Rapport historique sur les progrès des sciences mathématiques depuis 1789, et sur leur état actuel* (Paris, 1810), 264; *La grande encyclopédie, inventaire raisonné des sciences, des lettres et des arts*, ed. by André Berthelot and others (31 vols., Paris, 1886-1902), IX, 362. Figuier, *Merveilles*, IV, 34-35, asserts that there was a pumping lamp, run differently from Carcel's, already in operation in southern France and that Carcel was influenced by it.

lamp was destined to bring. Thus these French Aladdins of the eighteenth century had magic lamps that did not bring them gold.[17]

If Carcel had reached perfection—or near perfection—with the oil lamp, another Frenchman, Philippe Lebon (1769-1804) was to pave the way for lighting the world with a more brilliant lamp than any Carcel could manufacture. Lebon undertook to illuminate with artificial gas.

Philippe Lebon, the son of an officer in the king's household, was born at the little town of Bruchay, near Joinville, in what is today the department of Haute-Marne. He was given an excellent scientific education, first at Chalon-sur-Saône and later at the Ecole des Ponts et Chaussées, the notable engineering school founded by Trudaine and Perronet in the mid-century. He showed himself a brilliant student and on graduation received the rank of major. Sent first to Angoulême, near Bordeaux, as highway engineer, he was shortly recalled to Paris to teach mechanics at the Ecole des Ponts et Chaussées. For a time he labored at perfecting the steam engine and in 1792 was given a national award of 2,000 livres for improvements made on it.[18]

Around 1791 (Hoefer says 1797) he became interested, while on a visit to his father's home at Bruchay, in the extraction of gas from wood for the purposes of heating and lighting. His experimentation began when out of curiosity he placed some sawdust in a vial and held it over a flame. He observed that a gas was given off which caught fire as it came from the tube. It smoked badly. Lebon then tried to get rid of the substances that caused the smoke by running the gas through a water bath. This procedure did improve the quality of the gas, but it did not eliminate completely the smoke or resinous smell. On his return to Paris he discussed the problem with Fourcroy and others, to whom he told his dream of using the gas for heating and illumination.[19] These scientists all encouraged him to continue work on the

---

[17] Figuier, *Merveilles*, IV, 37-39.
[18] Tuetey, *Répertoire générale*, VII, 264.
[19] Figuier, *Merveilles*, IV, 97-99; *Nouvelle biographie générale*, XXIX, 105-106. Figuier says that at Bruchay, Lebon had exclaimed with delight before his neighbors that he would one day light both Paris and Bruchay. The neighbors simply shrugged their shoulders and muttered to one another, "He is a fool."

idea, which he did. Finally by the year VII (September, 1798–September, 1799) he had made sufficient progress to be able to read a paper on his experiments before the Institut de France. Later he applied for a *brevet d'invention* (patent) which was given him September 21, 1799. He called his lamp the *Thermolampe* (Heat Lamp), borrowing the terminology, as had others before him, from the Greek. He intended to use the lamp not only for lighting but also for heating a room. His work as engineer in Angoulême prevented him from devoting as much time to his experiments as he wished; at the same time, he experimented so much that he gained the dislike of his superior, who criticized him to the minister of the interior and tried to get him removed from his post. Lebon felt obliged to write the minister in his own behalf, telling of his promising experiments, of the encouragement he had received from noted scientists, and of the value the invention would be to France. The minister sided with Lebon and left him at his post, but paid him nothing even as he neglected to pay others because of the government's financial straits.

On November 30, 1800, Lebon tried to interest the government in his device for public heating and lighting, but without success. To demonstrate to the public the advantages of his *Thermolampe*, he leased the Hôtel Seignelay and for several months exhibited there a large one, which threw out an impressive amount of light and heat and attracted crowds of visitors. Lebon appears to have had a flair for the theatrical, and his illumination of a fountain brought admiration from the crowds. He also published a booklet of twelve quarto pages to announce the nature and merits of his invention. Unfortunately his gas, lacking complete purification, emitted "a fetid odor" that displeased the crowds, which decided that it was really not a practical invention.

Lebon was never able to free his lamp from this drawback, nor was he able to win public acceptance. In 1801 he took out another *brevet d'invention* on his method of distillation, and in 1803 the government granted him a lease on part of the pine forest of Rouvray, near Le Havre. Prior to this concession a commission of judges examined his work and reported to the government that his experiments had "measured up to and even surpassed the hopes of friends of the sciences and arts." He had not long been engaged at exploiting the forest of Rouvray when he was twice approached by the Russian princes Galatzin and Dolgorowski with a proposal to take his invention to Russia. He refused. The next year he was mysteriously killed on the Champs Elysées in Paris by assassins, who stabbed him thirteen times. He had been called to Paris to help prepare the city for the coronation of Napoleon, and his death occurred on the inauguration day. His widow was left in poverty and in debt to the government. Late in 1811, government attention was called to her needs, and she was given a small pension until her death in 1813. By that time the English had developed Lebon's ideas and were prepared to utilize gas for the illumination of factories and cities.[20]

The man to whom historians have commonly given credit for the invention of gas illumination is the Scottish-born William Murdock (1754-1839), who worked about the same time as and independently of Lebon. Murdock never attended a university but learned from his father the trade of a millwright and in 1777 went to London where he obtained employment with Watt and Boulton. Until 1830 he was employed by this firm, and then pensioned. In 1779 he was sent by the company to Redruth in Cornwall to maintain a number of Watt engines in that district, and he remained there until 1798. It was at Redruth that he experimented with success at lighting his room with coal gas. The date is not certain, for no record was made until 1808 and at that time much confusion on details existed among the eye witnesses.[21] But it is certain that late in 1795 and early in 1796 Murdock conducted some experiments at the Neath Abbey Iron Works, burning coal gas which was made in an iron retort and conducted some three or four feet through an iron pipe. The light was "strong and beautiful." He also lighted his room at Redruth with the gas, conducting it from a retort outside the house by means of a tube through his window to a burning point near the ceiling. Returning to London in 1799, he demonstrated his

[20] Figuier, *Merveilles*, IV, 104-108; *Nouvelle biographie générale*, XXIX, 106-107.
[21] The date commonly given is 1792, and in 1892 a centenary celebration was held at Redruth; nevertheless, R. B. Prosser, author of the article on Murdock in *D.N.B.*, XIII, 1222, thinks that 1792 was "much too early" and suggests 1794 or 1795.

new light at the Watt and Boulton Soho Works. In 1801 Gregory Watt, son of the famous James Watt, wrote his father from France telling of Lebon's experimentation and urging that efforts be made to obtain a patent for Murdock at once.

In 1803 gas illumination was installed in a portion of the Soho factory. In 1804 the home of G. A. Lee, a cotton manufacturer in Manchester, was provided with equipment for gas lighting, and later that year Lee and his partner Phillips "decided to light their mills with gas." From that time on, lighting in England by coal gas made rapid strides, though in the enterprise Murdock had little part and derived little, if any, financial reward. In 1809 Parliament was presented a bill to license the Gas Light and Coke Company, "in which Murdock had no part," and much debate arose. Watt and Boulton labored hard to protect Murdock's claim to the invention, and investigations into the history of Murdock's experimentation were made. Though Murdock did not profit financially from his invention, he received in 1808 the Rumford gold medal by the Royal Society. It so happened that coal gas was free from the disagreeable features of wood gas and was universally employed by those installing gas illumination. Because of this fact the world is more indebted to Murdock than to Lebon. Caprice of fortune favored Murdock's material, but after all caprice has played no small part in history.

CHAPTER V

Papermaking

UNTIL LATE IN THE EIGHTEENTH CENTURY, PAPER IN EUROPE and America was made almost exclusively from linen and cotton. Occasionally hemp and wool were used, apparently in mixture with either cotton or linen, but not in the better grades of paper. Paper also was made from asbestos. In China, where paper of unsurpassed quality was made, silkworm skins, bamboo, and other materials were commonly used, but eighteenth-century Europeans knew little of Chinese manufacture of this article and were not at all influenced by it. To keep up with the enormous output in the 1700's of books, pamphlets, periodicals, and newspapers, not to mention stationery for business and personal uses, paper mills sought great quantities of linen and cotton rags, and newspapers frequently urged their readers to save all their tatters.[1]

The rags were first washed and bleached, then placed in large receptacles or basins for trituration. In eighteenth-century France, trituration, or grinding of the material into paste, was done by two methods: mallets and lacerating cylinders, both forms turned by water power. The mallet or stamping machine was the older device, having been originated by the Spaniards in the late Middle Ages. It operated on the principle of the earliest grain mills,

[1] Dard Hunter, *Papermaking: The History and Technique of an Ancient Craft* (New York, 1943), 230; Charles Thomas Davis, *The Manufacture of Paper: Being a Description of the Various Processes for the Fabrication, Coloring, and Finishing of Every Kind of Paper* . . . (Philadelphia and London, 1886), 26-28, 33, 36; *Encyclopédie méthodique, ou par ordre de matières*, ed. by Charles Panckoucke (229 vols., Paris, 1782-1832), *Arts et Métiers*, V, 471-72, 481, 502.

The book by Hunter is a highly interesting account by the leading contemporary scholar on the history of paper; the account by Davis is more succinct but clear; while the long article in *Encyclopédie méthodique*, V, 463-592, by Nicolas Desmarest, a member of the Academy of Sciences and inspector general of manufactures in France in 1788, is vastly more detailed, the work of an eighteenth-century expert.

where a pestle crushed grain in a hollowed stone. A paper manufactory usually had several basins, each provided with a heavy mallet of wood or metal. Before stamping, the cloth was soaked a long time to soften it, but decomposition was avoided. Sometimes lime, used to speed the softening process, damaged the fiber, and the paper made from it was inferior in quality. Late in the eighteenth century the French government forbade its use in papermaking.[2]

The second triturating machine, the lacerating cylinder, had been invented, or at least perfected, in Holland in the seventeenth century. It was a log or metal bar with thirty or more sharp blades attached, which was rotated in a long basin or tub, cutting the rags to shreds against the stone or metal bottom. The French paper mills at Montargis used smaller cylinders than did the Dutch in the late 1700's (the latter weighed about 3,000 pounds each), and operated them at a greater speed, from 138 to 166 revolutions a minute in contrast to the 66 revolutions of the heavier Dutch ones. Nicolas Desmarest, a French papermaking authority, greatly preferred the cylinders to the mallets, partly because they turned out a smoother paste and partly because their work was much faster. Two cylinders, he stated, could perform as much work as eighty mallets and occupied vastly less space. He urged the French to abandon the mallets altogether in favor of the cylinders.[3] Moreover, the cylinders dispensed with the preliminary water-softening and also with the repeated washings needed to cleanse the paste after its pounding under the mallets.[4]

The macerated fiber then was taken in buckets to the dipping vats. In France these were wooden tubs, ordinarily five feet in diameter and two and a half feet deep. They were bound by three or four iron bands for reinforcement. The water in these vats was kept at a constant temperature by a coal or wood stove placed in a circular enclosure of metal in the vat. By the vat stood a workman who dipped into the liquid, fibrous material a rectangular wooden mold resembling a tray with a stick or handle

[2] Hunter, *Papermaking*, 112-18; *Encyclopédie méthodique*, V, 483-91. Hunter gives illustrations of these machines.

[3] *Encyclopédie méthodique*, V, 494-96.

[4] Hunter, *Papermaking*, 119-20, 124-25.

below. Over the sievelike bottom (made of little spruce spindles) was stretched a removable piece of cloth. The workman would carefully gather on the mold the proper amount of paste and permit the excess water to drain back into the vat. Then he would remove the cloth from the mold.

Over this cloth at the time of removal was placed a piece of felt, and the two were reversed in position, so that the felt now received the paste and drainage. Succeeding layers of paste and felt were added until a pile had accumulated, when they were subjected to pressure, first under a long lever and then under a screw press. This pressure not only removed the surplus water, but also gave a needed consistency to the fibrous paste.

The felts then were removed and the pressed sheets carried to a drying loft, where they were hung in "spurs" *(porses)* of five or six sheets each on horsehair ropes covered with beeswax. Air was admitted as desired by means of wooden shutters.[5]

After drying, the paper was taken to the sizing or glue room, where the spurs were dipped in a tub of warm glue made by boiling bits of skin obtained from tanners, leather dealers, and parchment makers. Before being used, the sizing was twice heated and twice strained through a cloth to free it from foreign matter. The dipped sheets, after draining, were pressed, separated, and hung to dry singly on cords before a heated stove. The paper dried rapidly. The room was closed so that all parts of the paper received the same heat and dried equally. Most of the size was lost from the paper through evaporation. Sizing had the double advantage of making the paper both firmer and smoother. Desmarest, commenting on the paper of his day, said that the French were not so careful to obtain as smooth a finish as were the Dutch, and that some of their paper was rough because the sheets were pulled apart after the size had been applied.

The dried paper was sorted, counted, and placed in lots *(mains)* of twenty-five sheets. These in turn were placed in reams *(rames).* French paper workers turned out more or less the same amount

---

[5] *Encyclopédie méthodique,* V, 496-516; Hunter, *Papermaking,* 133-42; Davis, *Manufacture of Paper,* 96-97. It may be added that felt was used for the expulsion of water because it is the only fibrous material to which the wet paste will not stick. G. S. Witham, Sr., *Modern Pulp and Paper Making: A Practical Treatise* (2d ed., New York, 1942), 424.

of work each day. They were jealous of their work and would not jeopardize quality for quantity.[6] One of the largest French mills in 1760 produced only about seventy-five reams a day. It was handmade paper of the highest quality, but made in laborious fashion and accordingly expensive.[7] Desmarest listed two or three dozen types of paper, each with its name, weight, cost, and other details presented in a chart.

Coarse paper made from the dregs of the vat was used for wrapping material. In addition, much paper was made faulty by the vatman's not removing knotty materials while molding the sheets. The French exported this coarser type.[8]

Most handmade paper of that day, as afterward, bore watermarks or impressed designs. These were made by placing a device on the mold when the paste was dipped from the vat.[9]

Mordants were used to color paper. A certain amount of alum was commonly placed in the sizing tubs, and some manufacturers added zinc sulphate (vitriol of zinc). Other mordants used later, and possibly even in the 1700's, included iron sulphate (green vitriol), lead nitrate, and lead acetate (sugar of lead).[10]

As the supply fell more and more below the demand of the paper mills, substitutes were sought. As early as 1716 a book in England suggested the manufacture of paper from raw hemp, but there is no evidence that it was attempted.[11]

In a treatise dated November 15, 1732, the famous French naturalist René Antoine Ferchault de Réaumur (1683-1757) proposed to the Royal Academy of Sciences that paper be made of wood. He called attention to the fact that in America wasps, especially those of Canada, made paper of a high quality out of wood fiber. He considered it possible for man also to make paper from wood and plants. He recognized that there was a difference in types of wood, and he was not certain whether Europe had

---

[6] *Encyclopédie méthodique*, V, 510, 517-24, 562; Hunter, *Papermaking*, 142, 148; Davis, *Manufacture of Paper*, 97-98.

[7] Hunter, *Papermaking*, 120.

[8] *Ibid.*; *Encyclopédie méthodique*, V, 482.

[9] Davis, *Manufacture of Paper*, 40-41, 98.

[10] *Ibid.*, 462; *Encyclopédie méthodique*, V, 518.

[11] Hunter, *Papermaking*, 233. The book was entitled *Essays, for the Month of December 1716, to be continued Monthly, by a Society of Gentlemen.* Essay VI dealt with this matter.

wood as suitable for paper of quality as did America. He observed that it would be necessary to break down this wood into soft paste, even as did the wasps. But whatever the difficulties to overcome, the ever-increasing use of paper by Europeans appeared to him to demand that this source be explored.

So far as historians know, Réaumur (1719) was the first to suggest the making of paper from wood, which today is the source of most of the world's newsprint and wrapping paper. Linen still forms the base of the most elegant paper, but this is only a small portion of the paper in use. Oddly enough, Réaumur never made the experiment that he recommended. In 1742 he remarked that though more than twenty years had elapsed since first the idea occurred to him, he regretted that he had never attempted its demonstration.[12]

His idea was shortly applied by several other men, although it is not clear that they owed their inspiration to him. In 1727-1730 the German Franz Ernst Brückmann printed on paper made from asbestos several copies of a two-volume work on geology. In 1741, Jean Etienne Guettard (1715-1786), physician to the Duke of Orleans and member of the Academy of Sciences, wrote several articles suggesting the use of swamp moss (conserva) for the manufacture of paper. He sent with one of the articles experimental samples of paper made from the leaves, bark, and wood of various trees and plants. His treatise was translated into English and published in London (1754) and Philadelphia (1777). In this essay he recommended also various vegetables as sources for paper.[13]

In a four-volume work published in 1734-1765 the Flemish naturalist Albert Seba suggested that seaweed and Muscovy mats be utilized for paper manufacture. The English traveler John Strange reported in a book written in 1764 that the town of Cortona, Italy, was engaged in the manufacture of paper from broom and other plants.[14]

[12] *Ibid.*, 232-34. The best biographical account of Réaumur is Jean Torlais, *Un esprit encyclopédique en dehors de "l'Encyclopédie": Réaumur d'après des documents inédits* (Paris, 1936).

[13] Hunter, *Papermaking*, 235-36. For a biographical sketch of Guettard, see M. J. A. N. Caritat, Marquis de Condorcet, *Oeuvres complètes* (21 vols., Brunswick and Paris, 1804), III, 317-47.

[14] Hunter, *Papermaking*, 236.

It remained, however, for a German clergyman named Jacob Christian Schaffer to demonstrate thoroughly to Europe the possibility of manufacture of paper from a long list of trees and plants, and even from other materials such as asbestos. Over a long period he made the study of paper his hobby and at length published a six-volume work on the manufacture of paper entitled *Versuche und Muster ohne alle Lumpen oder doch mit einem geringen Zusatze derselben Papier zu machen* (Regensburg, 1765-1771). In it he submitted eighty-one samples of paper made from various materials.[15] There could no longer be any doubt of the possibility of paper manufacture from bases other than linen and cotton. The glory might have gone exclusively to French genius, had Réaumur demonstrated his idea.

The Swedish chemist Karl Wilhelm Scheele (1742-1786), who in 1774 isolated chlorine gas and observed its bleaching tendencies, was indirectly responsible for a process of bleaching paper by the use of chlorine gas and lime, in solution.[16] As already noted, however, lime injured the fiber of the paper, and the French government forbade its use.

During the French Revolution, when the shortage of paper was acute (partly from its use as gun-wadding), newspapers had to be reground and processed, and there was a strong demand for bleaching materials which would remove the printer's ink. Several persons submitted to the government methods for achieving this end. That of a woman, the Citoyenne Masson, was adjudged the best by a committee of investigation, which recommended that the government make an award of 3,500 livres. Her bleaching agent was soda.[17]

Another French invention of the century deserving mention was the papier-maché or wet-mat stereotyping process, devised by the printer Claude Genoux in 1728.[18]

[15] *Ibid.*, 237-44, 397.
[16] *Ibid.*, 237. There is no indication that Scheele himself devised the process. The value of lime as a bleaching agent for fibrous material had long been known, but it was charged with injuring it and early in the 1700's Britain had forbidden its use. Joseph H. Park and Esther Glouberman, "The Importance of Chemical Developments in the Textile Industry During the Industrial Revolution," in *Journal of Chemical Education*, IX (1932), 1150.
[17] *Procès-verbaux du comité d'instruction publique*, III, xcv-xcvi.
[18] Hunter, *Papermaking*, 332. Not until 1812 was this process adopted in America. *Ibid.*, 352.

The greatest French contribution to paper manufacture, however, was the discovery of a machine method of making paper of a fixed width and unlimited length. This was tantamount to the introduction of large-scale manufacture of paper. It was the work of François Nicolas Louis Robert (1761-1828), a native of Paris, who was by turns a soldier, a printer, and a schoolteacher. At the age of fifteen he attempted to join the French army, but because of his frail constitution he was rejected. Four years later he renewed his attempt and was received into a battalion of artillery stationed at Calais. About a year later (1781) he was sent to Santo Domingo to fight the British. Afterward, at the age of twenty-eight, he was discharged and took a position with the widely-known publishing house of Didot. Tiring of a clerkship, he obtained intimate knowledge of the processes of handmade paper. As inspector of personnel he also became well acquainted with the vexing behavior of the workers. In disgust at their petty bickering and quarrels (rather than from a desire to produce cheaper paper), so it is reported, Robert with the approval and encouragement of his employer, Didot Saint-Léger, set about to find a mode of making paper with as few workmen as possible.[19]

Working in the Didot plant at Essonnes and provided with money and equipment by Didot Saint-Léger, Robert soon constructed a machine (1797) that did not work. Dejected, he resigned to become superintendent of a grain mill. But after six months he was back again, and in 1798 produced a successful machine. Accompanied by Didot, he went at once to the office of the minister of the interior, François de Neufchâteau, with two samples of paper made by the new machine and applied for a patent *(brevet)*. Robert's letter of application for the patent, dated September 9, 1798,[20] explained that the machine was a labor-saving device and could manufacture paper of any length desired. All manual labor of whatever nature was supplanted by it.

It was an epoch-making instrument, as significant as some English textile inventions of the eighteenth century. The French

[19] *Ibid.*, 258-59.
[20] Quoted in full *ibid.*, 260-61. A picture of this machine, as reconstructed, is given on p. 261.

Conservatory of Arts and Trades, which under the minister of the interior had jurisdiction over inventions, sent a skilled draftsman to Essonnes to draw up a detailed description of the new machine. He praised the machine as achieving all the claims which Robert had made for it.[21]

On further examination the machine was approved, and in 1799 the government granted Robert a patent for the customary period of fifteen years and made him an award of 3,000 francs. The machine was yet in a crude stage, and it was expected that Robert would work to improve it.[22]

The fundamental feature of the new invention was a band of "metallic cloth" or closely-woven brass or copper mesh which revolved over a horizontal frame provided with metal rollers turned by a hand crank. Robert hoped in time to use water power, but reportedly had not the monetary resources necessary for the experimentation. A cylinder with copper blades spread paste from the dipping vat in a thin layer upon the wire mesh. The mesh moved with an oscillating motion designed to shake surplus water from the paste into the vats beneath. From the mesh the paste was passed through two felt rollers for removal of further water, after which it was apparently taken elsewhere for drying. The new machine could make paper up to twelve feet in width and fifty feet in length, the width of the paper being controlled by deckles fastened on the wire at whatever spacing was desired.[23]

Robert hoped to realize a profit from his invention, but like so many other inventors he was to be disappointed. He tried a brief while, in partnership with Grandin, to exploit the invention, but without success, and in 1800 he sold his patent to Didot for 25,000 francs, to be paid in installments. The payments lagged, however, and in June, 1801, he retook his patent. Twice he applied to the French government for further financial assistance, and when it was not forthcoming went to England in the hope that he might obtain monetary support there. Meanwhile, Didot

[21] Quoted in full *ibid.*, 261-62.
[22] *Ibid.*, 264; Davis, *Manufacture of Paper*, 47; Ballot, *Machinisme*, 559. Davis gives the figure as 8,000 francs, but Ballot and Hunter are probably correct.
[23] Ballot, *Machinisme*, 559; Hunter, *Papermaking*, 263-64; Davis, *Manufacture of Paper*, 47.

Saint-Léger had gone to England in 1800 with the patent and a model of the invention, and with the assistance of his brother-in-law John Gamble and a skilled mechanic named Donkin worked at the machine's improvement. From the British government Didot obtained a patent in April, 1801. Thus it apparently mattered little to Didot that the French patent reverted to Robert. In 1803 a British patent was given to Gamble for improvements he had made to the machine. The next year Didot and Gamble sold their rights to the two brothers Henry and Sealy Fourdrinier, who became bankrupt in an endeavor to improve the machine further. Though in time the Fourdriniers lost control of the machine, it came to be known after them and is still called the Fourdrinier machine.[24]

Robert returned to France around 1807. He still held his patent rights there, and he entered into a partnership with Guillot, a papermaker at Mesnil, to exploit it. But the machine still had imperfections, and after losing 21,000 francs in trying to make his fortune with it, Guillot discontinued its use. Throughout all its early years the machine seemed to be a millstone about the necks of its owners. The reason is not hard to explain, for according to a twentieth-century authority, G. S. Witham, Sr., the highly intricate, complex Fourdrinier (or Robert) machine performs its service only when all its parts function well.

In 1811 Didot bought from one A. F. Berte the French patent rights incorporating the improvements made to the machine in England. It was now an excellent machine, and Didot exploited it in the years that followed.

As for restless, quarrelsome, unfortunate Robert, he found it necessary to resort to schoolteaching in the tiny town of Dreux for the last fifteen years or so of his life in order to eke out a living. There he died in poverty in 1828, one of France's greatest inventors.[25] His machine, however, has continued in use, and with many improvements it is the means by which most paper in the world today is manufactured.

---

[24] Ballot, *Machinisme*, 560; Davis, *Manufacture of Paper*, 48-49; Hunter, *Papermaking*, 264-65; H. A. Maddox, *Paper: Its History, Sources, and Manufacture* (6th ed., London, 1939), 8-9. An elaborate account of the Fourdrinier machine as it has been developed in time, with some excellent illustrations, may be found in Witham, *Pulp and Paper Making*, 371-507.

[25] Ballot, *Machinisme*, 558.

Chapter VI
Chemical Inventions

CHEMISTRY WAS STILL IN AN EARLY STAGE IN THE EIGHTeenth century. After 1770 rapid developments were made by a large number of brilliant workers, British, French, German, and Swedish. No country contributed more than did France, which furnished such men as Lavoisier, Berthollet, Macquer, Guyton de Morveau, Darcet, Fourcroy, Daubenton, Chaptal, Dutrone, Leblanc, Vauquelin, and Descroizilles. To France, too, belongs the credit for certain developments that might be called inventions—those practical applications of chemistry to industry, the household arts, and methods of warfare.

Well at the top of useful applications of chemical knowledge must be reckoned the use of chlorine as a bleaching agent. The gas had been isolated first in 1774 by the Swedish chemist Scheele, who called it "dephlogisticated marine acid air," and not until 1810 was it given, by Humphry Davy, the name "chlorine," taken from the Greek word *chloros* meaning "greenish-yellow." In 1785 an article by Claude Louis Berthollet (1748-1822) in the *Journal de Physique* revealed to the world the bleaching possibilities of the new gas.[1]

Until that time, bleaching of cloth required several months. Sour milk ordinarily was used, and the materials to be bleached had to be carried regularly into the sunlight and spread out over considerable space. The task was tedious and demanding. The new method, however, was vastly quicker.

There were, of course, some problems to be met: how best to bring the materials for bleaching into contact with the gas and how to remove the smell of chlorine from the bleached goods. About two decades later, these matters were solved by other chemists.

[1] Leonard A. Coles, *The Book of Chemical Discovery* (London, 1933), 244; Ballot, *Machinisme,* 329.

Berthollet's method was to put the chlorine in solution. He found that the gas united readily with potassium hydroxide, forming a solution which he called *eau de Javelle,* a name it still retains. Two chemists, Bonjour and Welter, had collaborated with Berthollet. Shortly they went, the former to Valenciennes, the latter to Lille, to prevail upon manufacturers to adopt the new process. Welter met with much success, but Bonjour found the bleachers of Valenciennes bitterly opposed to his ideas and spent two or three years there in fruitless efforts.[2]

Improvement in the process was made about 1791 by the Rouen chemist Descroizilles (1751-1825), a man of college training and broad interests.[3] He invented a machine, the *bertholli-mètre,* with which he could bleach on a large scale. Several manufacturers took an interest in the possibilities of his work and prevailed on him to open a bleaching establishment at Lescure-les-Rouen, "which quickly became the most important of all France." Descroizilles became a man of means.[4]

From France *eau de Javelle* was taken in 1796 to England where improvements were rapidly made. In 1799 Charles Tennant of Glasgow discovered that when chlorine was united with slaked lime a bleaching powder was formed. This quickly became an important commercial product, although Tennant was not able to exploit his invention since his patent was made useless when it was proven in court that slaked lime previously had been used as a bleaching agent.[5]

Berthollet's idea was given to the world freely, and both he and France were deprived of benefits that would have ensued

[2] Ballot, *Machinisme,* 329-30; Park, "Chemical Developments," 1146-52.

[3] Born at Dieppe of a long line of apothecaries, Descroizilles had attended the Oratorian college of the city and later had studied chemistry at Paris. Returning to Rouen he became an apothecary, a royal demonstrator in chemistry, a manufacturer, and a frequenter of a literary salon. During the Revolution he was imprisoned, and from the jail he sent the government a memoir on the large-scale manufacture of gunpowder. Too valuable a man to be left in prison, he was placed by the government, along with Chaptal, in charge of all the saltpeter making in France. Ballot, *Machinisme,* 329-30; C. Richard, "Les savants et le salpêtre en Normandie sous la Terreur," in *Révolution française,* LXXVI (1923), 233-34.

[4] He was later to distinguish himself by inventing the "eclipsing lighthouse" *(phare à éclipse).* Ballot, *Machinisme,* 330, 530; Richard, "Savants et salpêtre," 234.

[5] Coles, *Chemical Discovery,* 245; Park, "Chemical Developments," 1152.

could the process have been kept secret. Secrecy was probably impossible, however, in that day when there were no international patents or copyrights.

Who was this Berthollet? A Savoyard born near Annecy in 1748, he attended a college at Chambéry, and later for four years studied medicine at the University of Turin. Then he set out for Paris, carrying in his pocket a single letter of introduction, to the great physician of Voltaire and Frederick the Great, Tronchin, who took a fancy to the youth that came from near his native Geneva and recommended him to the Duke of Orleans as physician. From his arrival in Paris until the Revolution Berthollet appears to have been in the party of the Duke of Orleans, who was building an elaborate system of patronage and faction and who successfully pushed Berthollet for the Academy of Sciences in 1781 and for the superintendency of dyeing and the directorship of the Gobelins in 1784. These last posts brought to Berthollet's attention the importance of dyes, and in 1791 he published a book on the subject—probably the best treatise on dyeing then to be had. In the Revolution he was given an active part in providing France with gunpowder. When Napoleon conquered Italy (1796-1797) Berthollet was among the scientists, savants, and artists appointed to go there to select materials for transportation back to France. Later he went to Egypt with Napoleon, made the march to Acre, and was one of the handful that Napoleon brought back from Egypt with him.

Berthollet contributed to chemical invention not only through his use of chlorine for bleaching; he was also a perfector of explosives. Throughout the 1770's and 1780's France, as also England,[6] was greatly interested in the discovery of better methods for manufacturing gunpowder. Turgot, despite his role as humanitarian, was keenly interested in this while controller general, 1774-1776.[7] In the 1790's war with most of Europe cut off France

---

[6] See the article on Bishop Richard Watson in *D.N.B.*, XX, 936.

[7] See A. R. J. Turgot, *Oeuvres de Turgot et documents le concernant, avec biographie et notes*, ed. by Gustave Schelle (5 vols., Paris, 1913-1923), IV, 376-78. Turgot persuaded the king to give 6,000 livres in prizes for the three best papers on the means of quicker and cheaper manufacture of gunpowder. The king also agreed to furnish a commodius place at the Arsenal or elsewhere for the commissioners' testing of the plans submitted. The Academy of Sciences conducted the contest, begun in 1775. Thirty-

from foreign supplies of nitrate. In desperation she called upon all citizens to dig up the ground in their stables, sheepfolds, cellars, and building ruins and strain it for saltpeter, offering the price of 24 sous a pound for all that could be brought in. Municipalities were urged to set up places for the washing out and evaporation of the nitrate-bearing earth. Families were instructed how to go about the process.[8] Berthollet, Fourcroy, Vauquelin, Chaptal, and other chemists were called upon to direct this national effort for gunpowder. They experimented not only with gunpowder but with other explosives. Berthollet worked at the substitution of saltpeter for cannon powder by the more violent potassium chlorate.[9] But two violent explosions, in one of which the powder mill at Essonnes was blown up and five persons killed, led to the cessation of experiment (instead of search for better methods of control). Berthollet later experimented with the yet more powerful fulminating silver *(argent fulminant)*. Attempts to use it for firearms had little success due to lack of control over its tendency to detonate at the slightest shock or with a sudden rise in temperature.[10]

These were not the first or the last attempts of the French in the eighteenth century to find a more powerful explosive force. In the days of Louis XIV, Robert Boyle in Britain and Nicolas Lémery in France had worked at the task; under Louis XV (1774) Bayen, chief pharmacist for the French armies, made known mercury fulminate *(fulminate de mercure)* and its characteristics. To the end of the century there was intermittent experimentation with various fulminates. When the silver fulminate of Berthollet was found too violent, French chemists experimented with mixing it, potassium chlorate, potassium iodate, and other substances in turn with various combustible substances, but they had little success in controlling them. An Englishman named Howard,

---

eight papers were submitted, but for some reason it was decided to prolong the contest, and in 1782 sixty-six papers were entered. The first prize of 4,000 livres went to Thouvenel, powder commissioner at Nancy.

[8] Richard, "Savants et salpêtre," 232-33.

[9] Figuier, *Merveilles*, III, 473, says that this was in 1788 and that Fourcroy and Vauquelin worked with him.

[10] *Ibid.*, 474; *Nouvelle biographie générale*, V, 718.

THE "FLOATING BATTERIES" ATTACKING GIBRALTAR

These ten supposedly unsinkable ships of the Franco-Spanish fleet unsuccessfully bombarded the British fort in 1782.

(Illustration from Samuel Ancell, *A Journal of the Late and Important Blockade and Siege of Gibraltar*, facing p. 224, courtesy of the Library of Congress)

CHEMICAL INVENTIONS

working on the French idea, had better fortune in 1800, when he succeeded in making "an extremely explosive powder" capable of control, composed of mercurial fulminate and saltpeter.[11]

A third French chemical invention of this period was Nicolas Leblanc's method of making artificial soda from sea salt and sulphuric acid as raw materials (1789). The inventor (1742-1806), a pupil of Darcet and surgeon to the Duke of Orleans, had won a contest in 1787 sponsored by the Academy of Sciences for the manufacture of soda from marine salt. The Duke of Orleans set up a factory for him at Saint-Denis in 1787, but Leblanc met defeat and economic misery at the hands of soda importers who opposed his method.[12] He died in "abject poverty" in 1806, but his method, soon to become "world-famous," was taken to England where Muspratt erected a factory in 1823 and manufactured soda on a large scale. "For forty or fifty years it [this process] had no serious competitor; it was the only method of any importance for obtaining soda, a compound for which there was a rapidly increasing demand."[13] Once again the French failed to see and to exploit their great opportunities.

On another chemical invention—a new method for the manufacture of steel, by Réaumur—the French fared better. This invention has been described by the historian J. B. Perkins as one of the most useful developments for French welfare in the eighteenth century. Réaumur was one of the greatest entomologists of all time, but reputedly he knew little chemistry. It is therefore remarkable that he should have suggested two chemical inven-

[11] Figuier, *Merveilles*, III, 474; J. R. Partington, *A Short History of Chemistry* (London, 1939), 74.

[12] E. Levasseur, *Histoire des classes ouvrières et de l'industrie en France de 1789 à 1870* (2d ed., 2 vols., Paris, 1903-1904), I, 272-73, 411; Wolf, *History of Science*, 647-48. Leblanc's method consisted of decomposing marine salt with sulphuric acid. Sodium sulphate was obtained. This in turn was heated in a revolving furnace with large quantities of crushed limestone and coal. Carbon monoxide was given off in the heating, and the residue was a mixture of calcium sulphide and sodium carbonate, plus a number of impurities. This was treated with water, which dissolved the sodium carbonate and left the calcium sulphide. The solution was removed and evaporated, leaving impure sodium carbonate, which in turn had to be dissolved in hot water and allowed to cool. From this, sodium carbonate, or household soda, crystallized. Coles, *Chemical Discovery*, 222-23; Wolf, *History of Science*, 647.

[13] Coles, *Chemical Discovery*, 222-23; Wolf, *History of Science*, 648.

tions of first-rate importance, the manufacture of paper from wood and the manufacture of steel by an improved method.[14] Réaumur's process removed impurities in the iron by burning molten ore with sulphur in a closed container for a long period of time. If the resultant mass had too much sulphur and became hard and brittle, Réaumur would heat it again with lime to soften it. A company was created to exploit the new process, but it failed after a few years.[15]

Near the close of the century a superior form of steel, similar to that introduced in England about 1750 by Huntsman and Marshall, called crucible or cast steel, was invented independently by the academician Louis Clouet (1707-1801), a graduate of the engineering school at Mezières who had been working at the task since 1788. In their report of July 3, 1798, to the National Institute of the Sciences and Arts, Darcet and Guyton de Morveau gave it high praise and foresaw for it great industrial possibilities. Unfortunately the French neglected to exploit this remarkable invention. The secret of making this steel, according to the investigating chemists, was the use of carbonic acid rather than carbon with the molten iron.[16]

Réaumur served France greatly not only in his discovery of a practical mode of manufacturing steel but also in his discovery of the art of porcelain manufacture. For this he was in part indebted to a Jesuit missionary in the Far East who wrote him that the ingredients of Oriental porcelain were petuntse and kaolin and sent him samples. It was necessary, however, for Réaumur to experiment with these substances and with glass before he produced an excellent porcelain (1740) that made possible the celebrated manufactures of Sèvres and Limoges.[17]

---

[14] A long account of Réaumur's method (and also of the previous method of making steel in that time) is given by Joseph Bertrand, *L'Académie des Sciences et les académiciens de 1666 à 1793* (Paris, 1869), 378-80. Réaumur was "the first savant who did precise research on the nature of steel." Bertrand Gille, *Les origines de la grande industrie métallurgique en France* (Paris, 1947), 97.

[15] Bertrand, *Académie des Sciences*, 383-84. Bertrand recognizes the merit of the process but does not support Perkins on the matter of its value to France.

[16] *Mémoires de l'Institut national des Sciences et Arts. Sciences mathématiques et physiques, pour les ans IV-XIII* [1796-1805] (6 vols., Paris, 1798-1806), II, 81-97; Gille, *Industrie métallurgique*, 98.

[17] A full report of Réaumur's procedure is found in *Histoire de l'Académie,*

A chemical invention that brought large monetary returns was a lacquer varnish invented by Martin early in the century in imitation of the lacquer used on furniture by the Chinese. Throughout the late 1600's and the whole of the 1700's Western Europe bought heavily of Oriental wares, and this taste for things Oriental came to play a dominating part in the industrial and fine arts of the period. The mania in Britain for lacquered furniture prevailed also in France. Martin was one of those Europeans who succeeded in devising a European imitation of Oriental luxury goods. His varnish came to be made in all colors and for various types of objects—furniture, room and carriage decorations, toys, and above all else, snuffboxes. At first it was often as expensive as the objects that it covered, but considered of exquisite beauty, it was quite the fashion. First appearing during the Regency, it was still the vogue throughout the reigns of Louis XV and Louis XVI. Louis XVI declared its manufacture and sale a national enterprise and gave it monopolistic rights in France. Even in England as late as 1763, no other lacquer could compete with it either in quality or in sales. It was "a lacquer not inferior to that of the Orient itself."[18] Unfortunately the composition of this remarkable lacquer is now unknown.

A notable invention was the discovery in 1749 by Jean Jacques Bachelier (1724-1805) of the ancients' method of encaustic painting (painting in colors mixed with wax, burned in when the wax was not liquid). In this he anticipated by several years the Marquis de Caylus, who in 1754 by independent effort arrived at the same end. Bachelier proceeded to paint in encaustic a picture entitled "Flora and Zephyr." Apparently he did not apply to the government at the time for recognition of his discovery, for the first official notice of it seems to have been in 1792, when the Legislative Assembly made him a gift of 4,000 livres as "inven-

---

1740, pp. 56-58. Réaumur also invented around 1730 the thermometer named after him, in common usage in France after 1750. Its scale showed eighty degrees between freezing and boiling points of water. Other thermometers preceded Réaumur's, of course.

[18] Lacroix, *XVIIIme siècle*, 469-70,

519; B. Sprague Allen, *Tides in English Taste (1619-1800): A Background for Study of Literature* (2 vols., Cambridge, Mass., 1937), I, 205. In 1763 the English Society for the Encouragement of Arts, Manufactures, and Commerce offered a prize of 15 guineas for a lacquer to match it. The prize was won by Stephen Bedford.

tor of a wax varnish that he applies to oil paintings, which prevents harmful effects to these paintings." The government of the Old Regime, indeed, had given him a pension of 600 livres in 1778, but on the grounds of his being a royal painter and member of the Academy of Painting.

Bachelier had other claims to fame. For forty-four years he was director of the porcelain factory at Sèvres, and as such changed the decorations on porcelain from Chinese to French settings. He particularly stressed the "gallant fêtes." He founded in 1763 a Free Drawing School in Paris, with an outlay of 60,000 livres of his own, and despite sharp attacks from some quarters it became so popular as to have 1,500 students in 1766. He is credited moreover with devising a wax or encaustic for preventing marble statues and even certain lichens from deteriorating.[19]

A superior form of gilding, reported equal to the best produced in China and Japan, was devised by the chemist Torin and approved as an invention by the Academy of Sciences in 1767. Its secret was a superior mordant applied before gold leaf, making the gilding "much more brilliant and beautiful than that made with other known mordants" and also apparently more solid.[20]

Early in the century a certain Jannin of Paris invented a formula for making artificial pearls. So beautiful were they that according to Réaumur, who read a paper on them before the Academy of Sciences about 1719, they could deceive even an expert jeweller. They already had become fashionable, and Jannin had to employ "a large number of workers" to be able to supply the market.[21]

A superior form of lead pencil was invented in 1795 by Nicolas Jacques Conté (1755-1805), already mentioned in connection with the early Ballooning Corps of the French army. The out-

---

[19] *Grand dictionnaire universel du XIXe siècle* . . . , ed. by Pierre Larousse (17 vols., Paris, 1865-1890), II, 30; *ibid.*, VII, 494; *Archives parlementaires*, XIV, 17; Tuetey, *Répertoire générale*, VII, 265-66; *Nouvelle biographie générale*, IV, 57-58. In December, 1793, the claim of a certain Picault to the invention of a secret process for restoring and preserving paintings was reported to the convention, but the findings of the investigation committee are not clear. *Archives parlementaires*, LXXXI, 97.

[20] *Histoire de l'Académie, 1767*, p. 185. Torin later devised a means of converting peat to charcoal and was rewarded by the government in 1792 with a grant of 2,000 livres. Tuetey, *Répertoire générale*, VII, 263.

[21] Torlais, *Esprit encyclopédique*, 51.

break of war between Britain and France in 1793 deprived France of plombagine, which came from northern England. Knowing of Conté's resourcefulness, Lazare Carnot, the French minister of war, called upon him to solve the problem. This Conté did in eight days, devising a new substance—a mixture of pulverized graphite and argil (potter's clay) heated in a crucible and molded. The tiny molded cylinders were then placed in the grooves of cedar staffs, these being semicircular at the moment of the lead's injection and then sealed with glue to form pencils like those of the present day. The graphite mixtures could be made also in a hard form for drawing. The committee reporting on the pencils to the Academy of Sciences stated that already they had been made commercially for two years and had met every test with success. It declared them superior to the English product and predicted that they would render France independent of English and German pencils.[22]

Conté was not only an inventor of graphite pencils and a method of supplying balloons with gas, but also of a tool for coining money and a process for bleaching cloth superior to any in use at the time.[23]

This brilliant man was of peasant descent, born in the little village of Saint-Cénari in lower Normandy. His mother was soon left a widow with five other children to rear. Happily, two aunts who had entered a religious order took an interest in Conté and obtained their superior's permission to keep and teach the child. He early showed signs of brilliance, at the age of nine carving a violin which impressed craftsmen in that line. At thirteen he undertook to paint a chapel when the artist engaged at the task fell ill. Such praise did he win that he decided to become a painter; he went to Paris and took lessons under Greuze; and finally in 1779, after three years' training, returned to Sées to

[22] *Mémoires de l'Institut*, II, 98-106; Maurice d'Ocagne, "Un inventeur oublié: N.-J. Conté," in *Revue des deux mondes*, 8th ser., XXII (1934), 917-18. The committee of the Institut, Bayen and Fourcroy, reported that in 1771 the Academy of Sciences had recognized a pencil by Lafosse, a skillful engraver, but that he had never commercialized his process; that Bachelier, the famous painter, likewise had invented a drawing pencil for his own use; and that Desmarais had received from the Bureau of Consultation an award for a pencil of his invention. It remained for Conté, however, to provide a pencil for the public.

[23] Ocagne, "Conté," 916-17.

become a popular portrait painter. In 1785 he went to Paris with his wife and daughter, his wife having married him despite her noble family's opposition. Among the people whose portraits he painted in Paris was the Duchess d'Orleans, mother of Louis Philippe.

While plying his art in the capital, Conté became interested in science. The invention of the balloon intoxicated him. He spent evenings devouring books on the sciences. Soon he met men of distinction in science—Leroy, Charles, Vandermonde, Guyton de Morveau, Fourcroy, and Vauquelin—and gained their respect. The Revolution led him to abandon painting for science. He had an exciting and adventurous career in France, Germany, and Egypt. Napoleon praised his services in Egypt and later made him one of the first members of the Legion of Honor. Unfortunately, his last years were marked by frail health, and he died in 1805.[24]

Other chemical inventions by the French included the extraction of gelatine from bones by Darcet (1727-1801) and the distillation of sea water for drinking purposes by Poissonnier (1720-1798), a noted physician who for a time was consultant to Catherine the Great. Poissonnier was anticipated about two decades by Stephen Hales of England, but he was given credit for an independent invention. His method was used with success by Bougainville on his long voyage around the world (1766-1769).[25] J. B. M. Meusnier (1754-1793), the brilliant army engineer who worked at the perfection of the lamp wick, improved the mode of distilling sea water and devised a bouillon tablet for serving the sick on vessels. For this latter the government gave him pensions in 1777 and 1781 totaling 1,000 livres annually.[26]

A better and cheaper method of purifying cane sugar was made by the chemist Dutrone about the year 1795, for which the government rewarded him in 1792 with 6,000 livres.[27] French

[24] Ibid., 912-24.
[25] Levasseur, Classes ouvrières, I, 412; Mercure historique et politique (200 vols., Parma and The Hague, 1686-1782), CLVI (1764), 283-86; Archives parlementaires, XIV, 292.
[26] Archives parlementaires, XIV, 236; Bondois, "Industrie et commerce," 181-82.
[27] Ibid., VII, 269.

scientists were also working in the 1790's at the task of extracting sugar from the sugar beet where its presence had been demonstrated around 1747 by the noted Prussian chemist Margraff (1709-1780). Late in the century, Achard of Berlin and Deyeux and Parmentier of France worked on this problem, with Achard meeting the most success; but his results were long regarded as uneconomical.[28] Pierre Figuier, uncle of the author of *Merveilles de la Science*, also experimented with the sugar beet, being primarily interested in the bleaching of sugar. This he did by means of bones *(charbon animal)*.[29] In consequence of these and other developments, Napoleon became interested around 1811 in making France independent of foreign sugar, and ordered large tracts of land planted in beets.[30]

Another Frenchman, Nicolas Appert, was working at the turn of the century on the preservation of foodstuffs by canning. The ancients had known of the possibility of preserving fruits by means of honey.[31] In the mid-eighteenth century the noted English physician James Lind discovered that he could preserve fruit and vegetable juices in bottles by pouring olive oil over them and sealing the containers with cork. He also found it possible to preserve certain fresh vegetables for weeks by covering them with a layer of salt, which he later washed out before cooking.[32] When and how Appert began his work is not clear, but he met success in 1809 and introduced one of the most useful of modern inventions—an invention which was to have a notable part in freeing his country and the world from famine. Throughout the 1700's there had been few years in which some part of France did not suffer from famine or scarcity of food; subsequent French history has been different. In this metamorphosis, canning and improved means of transportation and communication

---

[28] J. A. Chaptal de Chanteloup, *Chemistry Applied to Arts and Manufactures* (4 vols., London, 1807), II, 393-94.

[29] Figuier, *Merveilles*, III, 474-75; Levasseur, *Classes ouvrières*, I, 413. Levasseur says that bones were used as a bleaching agent during the century, but he does not say for what.

[30] Henri Sée, *Histoire économique de la France: les temps modernes (1789-1914)* (Paris, 1942), 90-91.

[31] Well known is the story of Herod the Great's preserving the body of his wife Mariamne in a glass coffin filled with honey.

[32] See the account in Hans Zinsser, *Rats, Lice and History: Being a Study in Biography* . . . (Boston, 1935), 288-89.

have played major roles. Canning was not an eighteenth-century invention, yet the experimentation leading to its discovery began in the 1700's.

Tremendous developments in the art of tanning leather were made in England during the last decade of the eighteenth century, but France, despite the great amount of tanning undertaken within her borders, played only a feeble part in this advance. That part was chiefly the work of Armand Seguin, who in 1794 or 1795 arrived at a shortening of the long, disagreeable methods of tanning then in use by putting the leather for a mere day or two in a solution containing dilute sulphuric acid and wood bark. Coming at the time it did, with France at war, it was a great military asset in speeding up the manufacture of boots and other leather equipment. For his service the government in 1795 gave Seguin an award.[33]

Although the French did little in the realm of tanning, they made remarkable strides in the field of dyeing. This development came almost wholly within the latter part of the eighteenth century and was the result in part of certain fortuitous discoveries and thefts of secrets, and in part of the solid work of a number of able chemists.

The two men who contributed most in this field were Pierre Joseph Macquer (1718-1784) and Berthollet. Macquer was born in Paris of a Scottish family devoted to the Jacobite cause. He was educated at the University of Paris, where in 1742 he received his doctorate in medicine. Though he never lost his interest in medicine and later became one of the first members of the Royal Society of Medicine (founded in 1776), he seems to have practiced it little, if at all. Almost immediately on his graduation he set out on the avocation of chemistry. His progress was rapid. In 1745 he wrote his first paper in that field and was

---

[33] Paul Baud, *L'industrie chimique en France: étude historique et géographique* (Paris, 1932), 123; Rambaud, "Sciences," 110. Baud has a long chapter on the French tanning industry in the 1700's. Chaptal also has a chapter of interest on this matter. It appears from Chaptal, *Chemistry Applied*, IV, 235, that David MacBride in 1774 or 1778 had suggested the use of a small quantity of sulphuric acid for tanning hides. Either Seguin had not heard of MacBride's work or he did not deserve the credit of the invention. *D.N.B.*, XII, 425, says that MacBride employed lime water in tanning as early as 1767, but says nothing of his using sulphuric acid.

chosen an *adjoint* (assistant) in the chemical division of the Royal Academy of Sciences. By 1748 he had made a substantial contribution to knowledge of dyes, and two years later the government appointed him inspector of inventions and discoveries in chemistry. Shortly afterward he was chosen professor of chemistry at the Jardin du Roi. He became assistant, and later successor, to the chemist Hellot as Director of the Dyeing Industries and Director of the Royal Factory at Sèvres (in porcelain), in which posts he made important contributions. He was a member of several foreign academies—Madrid, Turin, Stockholm, and Philadelphia. He published many writings in the field of chemistry, one of which was his *Dictionnaire de chymie* (1778), the first chemical dictionary in any language. He lived at a time when developments in chemistry were coming rapidly, and he kept informed of them and communicated them to the public.[34]

Macquer rendered notable service in explaining scientifically the role of mordants in dyeing. The world already was familiar with the use of mordants to provide a foundation to which the dye could cling, but did not understand why not all dyes or types of fabric required mordants or why, when mordants were needed, they could in some instances be mixed with the dye but in others had to be applied in advance.[35]

Through experimentation with mordants, Macquer came to discover the means of applying certain new dyes. One was a beautiful red for silks, obtained by first dipping the silk in a mordant bath of tin compound and then in a bath of cochineal.[36] Another was Prussian blue, a color discovered in 1704 by a Berlin dye manufacturer named Diesbach, who subsequently sold

[34] L. J. M. Coleby, *The Chemical Studies of P. J. Macquer* (London, 1938), 9-15.

[35] *Ibid.*, 93-95; *Larousse du XXe siècle*, ed. by Paul Augé (6 vols., Paris, 1928-1933), IV, 982. The most common mordant of that period was alum. In the 1700's it was commonly found in an impure state. Chaptal, the French chemist, devised a method for purifying it. Levasseur, *Classes ouvrières*, I, 411 n. 4. Another mordant, especially used with cochineal, was a compound of tin.

In the nineteenth century, biologists wrestled with similar problems in perfecting staining techniques applicable to microscopy.

[36] Prior to Macquer's day, wool had been dyed by means of cochineal mixed with a mordant solution of tin compound, but it was not possible to dye silk in this fashion. Silk would not take cochineal, and it was left to Macquer to explain the reason. Coleby, *Chemical Studies of Macquer*, 90-91.

it to painters. As a paint, Prussian blue was well known in Macquer's day, but all attempts to use it as a dye proved disappointing because it would fade badly after two or three weeks. After careful analysis, Macquer discovered the reason for the fading. By first boiling his cloth in a mordant of alum and iron sulphate, then placing it for a minute in an almost boiling solution of alkali "calcined with organic matter," thirdly removing it to the open air for drying, and finally dipping it in boiling water containing a faint solution of sulphuric acid, he was able to dye it a beautiful, permanent bright blue. This discovery he made in 1748, describing it in a paper published in the *Mémoires* of the Royal Academy of Sciences in 1749. He found, moreover, that this color could be deepened by successive dippings, even to six times. On each occasion the cloth had to be dipped afresh in the alkali and the acid, which did not harm the cloth.[37] Prussian blue became an important dye. By 1770 the English had learned its formula and a company at Newcastle was manufacturing it. Many other colors or shades were produced by Macquer through his experimentation with mordants. His biographer reports that in 1777 "he presented to the Academy a series of more than forty new shades, all fast, which he had obtained with various dyes by using a solution of tin as his mordant."[38]

Berthollet, a student of Macquer, succeeded him in his position as Director of Dyeing Industries and carried on his tradition. He himself made no inventions or discoveries in the field of dyes, but he contributed greatly by a treatise, *Elements de l'art de la teinturie* (first published in 1791 and subsequently expanded in 1804 into two volumes), in which he explained the chemical action in dyeing. This excellent work embodied the most complete information on dyes in its day, and for a half century or longer had much influence in Europe.[39]

While these two remarkable men put dyeing on a scientific basis and paved the way for its future development, there were

---

[37] It was found necessary to increase or renew the alkaline solution on each dipping. *Ibid.*, 52-58, 85-88.

[38] *Ibid.*, 95; Park, "Chemical Development," 1158.

[39] Coleby, *Chemical Studies of Macquer*, 95-96; Kiréevsky, *Histoire des législateurs chimistes: Lavoisier, Berthollet, Humphry Davy* (Frankfurt a. M., 1845), 90-91, 106.

many interesting discoveries (in some instances no doubt thefts) by lesser lights. Several of them concerned "Adrianople red." This color, so fashionable in opulent French circles, had in the earlier half of the century been obtainable only from Turkey. In 1746 its secret was discovered by François Goudar, a weaver of fine cloths at Aubenas in Languedoc, who reportedly obtained it from the Greeks.[40] Goudar received financial aid from the Estates of Languedoc and the privilege, granted August 26, 1747, of dyeing cotton, linen, and hemp with the red or crimson of Adrianople.[41] As a result, his manufactory at Aubenas rapidly came to be one of the most successful in France. His prosperity led to the "discovery" of the secret by a merchant Eymar of nearby Nîmes, who likewise appealed to the Estates of Languedoc and obtained from them a grant of 10,000 livres. From Paris came an order of council dated November 22, 1757, permitting him "to set up an establishment wherever he desired for dyeing in Adrianople red." He added a textile mill to his dyeing works, and down to the Revolution, government inspectors reported that the quality of his dyes was superb, being even more beautiful and solid than those of Adrianople or India.[42]

In 1767 a third person in Languedoc, named Chauvet, came forward with the claim that he, too, had discovered the secret of "Adrianople red" and asked the Estates of the province to aid him in his dyeing establishment at Castelnau; but the Estates considered that in Goudar and Eymar they had shown enough encouragement of this dye. Chauvet in turn was followed by other claimants, notably from Montpellier.[43]

[40] Léon Dutil, *L'état économique de Languedoc à la fin de l'ancien régime (1750-1789)* (Paris, 1911), 504, 518. The process of dyeing "Adrianople red" was lengthy and complicated. According to Chaptal, writing at the close of the century, "A month was scarcely sufficient for completing all the operations that one has judged indispensable to obtain a beautiful red called that of Adrianople, and one employs successively to get it soda, oil, gallnut, sulphate of alumina [alum], sumac, blood, gastric juice, madder, soap, nitro-muriate of tin, etc." Of these, oil, gallnut, and alum were essential as mordants. Chaptal describes their several functions graphically in an article in *Mémoires de l'Institut*, II, 288-94.

[41] Dutil, *Etat économique de Languedoc*, 518; Claude de Vic and J. Vaissete, *Histoire générale de Languedoc* . . . (16 vols., Toulouse, 1872-1904), XIII, 82.

[42] Dutil, *Etat économique de Languedoc*, 518.

[43] *Ibid.*, 518-19.

In other parts of France the story was in part repeated. About the same time that Eymar made his discovery in Languedoc, Hellot, chemist and Director of the Dyeing Industries in France, came to the same discovery and sold it to the Estates of Brittany.[44] Similarly, a lace merchant of Dernatal, near Rouen, made claim to it and on January 10, 1761, obtained from the council of finances at Paris a license to dye goods "Adrianople red." He prospered, and by the time of his death in 1781 his firm (for he had admitted into partnership with him two Rouennese financiers) had acquired profits of more than 100,000 écus (300,000 livres).[45] Still later, in 1791, Joseph Xavier Chabot reported to the committee of agriculture and commerce of the National Assembly that he had discovered a way to make the "red incarnate of Adrianople and the rose red of Smyrna."[46] Of course all of these claimants did not come independently to this discovery. Likely, as Léon Dutil suggests, most of them acquired knowledge of the formula by purchase or other secondary means. From France or Greece (accounts vary) the secret process was taken to England, where it was equally popular. For obtaining the formula, the London Society of Arts in 1765 awarded £100 to Spurrit of Isleworth.

After the red dye of Adrianople, the "red of the Indies" was next most prized. Discovery of a process for obtaining it was claimed on the eve of the Revolution by a citizen of Rennes named Delorme, who applied to the government for the privilege of exploiting it.[47]

The widow Pallouis, who for several years prior to the Revolution ran at Versailles and Compiègne profitable manufacturing establishments making a type of silk known as *soie vrai galette* or *galette réelle* in imitation of a Swiss pattern, discovered a process for dyeing silk a beautiful scarlet. Up to this time chemists had

---

[44] Lacroix, XVIII*me* siècle, 537.
[45] Alexandre Tuetey (ed.), *L'assistance publique à Paris pendant la Révolution* (4 vols., Paris, 1895-1897), II, 306, 309.
[46] Tuetey, *Répertoire générale*, III, 544.
[47] *Inventaire-sommaire des archives départementales antérieures à 1790.*

Ille-et-Vilaine, ed. by Edouard Quesnet and Paul Parfouru (Rennes, 1892), C 1554 (hereafter, *Archives de l'Ille-et-Vilaine*). According to Lacroix, XVIII*me* siècle, 537, however, it would appear that before this Oberkampf was reproducing the dyes of the Indies and Persia with great success in his factories at Jouy.

been able to dye silk various other colors, such as purple, rose, red, and crimson, but not scarlet. This discovery contributed in no small degree to the success of Madame Pallouis's manufactures.[48]

Better methods of dyeing in black were claimed, with apparent validity, by the Abbé Mazéas and François Merlin before the Revolution;[49] and in 1790-1791 by Dimo Stephanopoli, a retired surgeon of the military hospitals in Corsica. Before 1790 the black dye in French usage was obtained from gallnuts; Stephanopoli derived his dye from the bark of oak trees. It was pronounced better not only by the manufacturers and hat dyers of Paris, but also by the Academy of Sciences and the College of Pharmacy, to whom the matter was referred.[50] It was expected that hatters would benefit most from his findings.

These discoveries in dyes did not complete the list of cloth-coloring inventions. About 1770 a process of giving a soft, elegant finish to dyed wool was discovered by Antoine Albert, a physician and member of the academies of Toulouse and Montpellier. English woolen goods previously excelled in this respect, but now French woolens so treated, whether in red, blue, green, or scarlet, won priority over English woolens in Near Eastern markets and produced a fortune for manufacturers in Provence and Languedoc.[51]

The dyeing industry in France was aided enormously by the establishment of a number of important textile plants in the 1750's and 1760's, certain of them by enterprising foreigners, of whom Christophe Philippe Oberkampf, founder of the great mill at Jouy, near Versailles, in 1759, is the best known. The industry was aided in no less marked degree by government interest and

[48] Tuetey, *Assistance publique*, II, 230-37.
[49] Lacroix, *XVIII*me *siècle*, 537; *Archives parlementaires*, XIV, 235. Merlin was given a pension of 600 livres in 1783 on the ground that he had discovered a new means of dyeing silk black. Lacroix also gives the Abbé Mazéas credit for the discovery of a beautiful red dye.
[50] Tuetey, *Répertoire générale*, III, 544; *Procès-verbaux du comité d'instruction publique*, II, 138 n. 3, 140. He was given a recompense of 10,000 livres by the government: 6,000 because of this invention and 4,000 because of his age.
[51] Lacroix, *XVIII*me *siècle*, 534-35. In 1748 Albert was given a pension of 590 livres by the government for the discovery of a red dye of madder root, without using brazil wood as previously. *Archives parlementaires*, XIV, 352.

patronage. The government made appreciable grants to inventors of new shades in dyes and encouraged the growth of the madder root, used widely in obtaining certain red dyes.[52] With this aid, France led European nations in dyeing at the opening of the nineteenth century.

Last but not least among French chemical inventions of the 1700's was the creation of a nomenclature which is used to the present day. This was formulated and published in 1787 by a commission of French chemists—Lavoisier, Guyton de Morveau, Fourcroy, and Berthollet. Macquer and Guyton de Morveau had called attention to the need of such action. The discovery of the elements oxygen, hydrogen, and chlorine had imparted a revolutionary atmosphere to the chemists; the awkward nomenclature already in use seemed too clumsy; and the tables of Geoffroy, drawn up early in the century, though seen to be inaccurate, had nevertheless demonstrated that in chemical combinations and reactions the ratio of union or change could be expressed in low and constant figures. This paved the way for simpler nomenclature bearing Latin and Greek prefixes and suffixes.[53]

The committee discarded the current chemical terminology, which was lengthy and vague, and in its place chose names which bore a rational description of the substances indicated. As their basic step, they distinguished between elements and compounds. Elements were described by a single word, largely based on Latin or Greek derivatives, as "oxygen," "hydrogen," and "sulphur." Compounds were given double names indicating the elements composing them. Acids were called by such terms as "sulphuric acid," "nitric acid," and "carbonic acid." Distinction was drawn between "sulphuric acid" and "sulphurous acid," the former containing a larger quantity of oxygen. A similar distinction was made between the salts in regard to their oxygen content, "chlorates" containing more oxygen than "chlorites," and "sulphates" more than "sulphites." The new terminology enabled chemists to tell at a glance the exact nature of a substance, and was of enormous

---

[52] Baud, *Industrie chimique*, 100-101; Coleby, *Chemical Studies of Macquer*, 95-96.

[53] Coleby, *Chemical Studies of Macquer*, 15; F. J. Moore, *A History of Chemistry* (3d ed., New York, 1939), 56-58, 97. Lavoisier had helped pave the way by a table of elements. See Moore, *History of Chemistry*, 98-100.

service both as a timesaver and clarifier in setting forth chemical affinity. Fourcroy (1755-1809) did much to popularize it in his lectures and in his books. Chemists in other countries, save for a few conservatives, quickly followed suit, and the new nomenclature was universally adopted. Although some minor changes have since been made, it remains "the basis of that in use today."[54] It was a contribution in some respects comparable to the metric system, formulated in that same period by French scientists, and like it has been of incalculable service to men of science.

In the enormous development of chemistry during the eighteenth century, when it was transformed from one of the least developed to one of the well-advanced sciences, French chemists had an important part, and their activity in applied science closely followed that in pure science. The inventions and discoveries by French chemists in metallurgy, bleaching, dyeing, and the manufacture of other necessities of life were numerous and important. Indeed, one of the most striking differences between the inventions of the first half of the century and those of the second half was the remarkable number of a chemical character in the latter period.

[54] Wolf, *History of Science*, 383-86; Partington, *History of Chemistry*, 133-34.

CHAPTER VII

Textiles

TEXTILE INVENTIONS AND DEVELOPMENTS, LIKE THOSE IN dyes, were numerous in France during the 1700's, more especially during the last decades. Their story has been badly neglected, and the records all too commonly are but scanty references. In many instances it is not possible to understand clearly the features that led officials of the government to designate them as "inventions." Often they must have been what today would be called improvements.

The textile inventions were all concerned with one or more of three processes: carding or preparation of materials, spinning, and weaving. In the carding of cotton and wool, a dozen inventions, real and alleged, were made in the 1780's and 1790's, of which the most notable were those by Antoine Germondy (1782), Fournier des Granges (1783), Sarrazin (1791), Louis Martin (1796-1797), Tellié (ca. 1797), and Lhomond (1792).[1]

The carding machine of Antoine Germondy at Lyons, moved either by arm or water power, could prepare a hundred pounds of cotton a day. Two years later, the mule-powered invention of Simon Pla had an output of ten livres an hour. Because of the length of a workday at that time, it must have had an even larger daily production than Germondy's machine. That by Sarrazin was for the carding of wool and transmutation of it into felt. It prepared forty-eight pounds a day, a task that normally required the labor of eight women. A fairly expensive machine, it cost 800 to 900 livres. Sarrazin used nine such machines in his establishment employing more than 1,200 persons. The machine of Louis Martin was able to card only twenty-five to thirty pounds of wool for a workday of twelve hours, but the quality of its product was

[1] Ballot, *Machinisme*, 55-56, 179-80, 209-10, 265; Tuetey, *Répertoire générale*, III, 540; *ibid.*, VII, 265.

superior to the wool carded by hand. The machine by Tellié, likewise for carding wool, was an adaptation of carding machines for cotton. The inventor, who had several inventions to his credit, received a government award of 1,000 francs and was promised 200 francs for each of the first ten machines introduced into industry during the next two years. Most excellent of the cotton carding machines was that of Lhomond, the Paris physicist, which was officially reported after careful examination superior to all others, whether French or English, and which promised to make France independent of foreign models for carding and spinning.[2] This claim was strong, for since 1748 the English had invented three carding machines using the cylinder process. The last, by Richard Arkwright (about 1775), was so cleverly devised that it is reported to have been changed little down to the twentieth century.

A superior method of refining hemp was devised some years prior to the Revolution by one Grandville of Burgundy, who reputedly spent his life and fortune at the task. The government granted recognition to his invention, permitted him to exploit it economically, and after his death rewarded his wife and daughter with a pension of 1,200 livres.[3]

In spinning, the French anticipated the "spinning jenny" of James Hargreaves (1765) with two machines. One, approved by the Academy of Sciences in 1745, was by a certain André l'aîné of St. Jean-en-Royans, Dauphiné. The examiners, Hellot and De Montigny, reported that the machine spun cotton, flax, wool, and hemp. It consisted of two wheels, connected by a cord or belt and turned by a handle, and three posts, each with a bobbin. It spun three threads simultaneously instead of the usual one. This feature of multiple spinning was, of course, the characteristic of Hargreaves' eight-spindled invention, which came many years later.[4] Without question, André's machine was practical, but if it was ever used there is no mention.

[2] Tuetey, *Répertoire générale*, VII, 265; Waldemar Kaempffert, *A Popular History of American Invention* (2 vols., New York and London, 1924), II, 209-11.

[3] Tuetey, *Assistance publique*, II, 576-78. For several years after 1785 Grandville's process was used by the Bureau de Filature, a Paris institution for assisting indigents.

[4] An illustration and a detailed account of the instrument of André l'aîné

The second machine embodying the principle of multiple spinning was invented in 1755 by François Nicolas Brizout de Barneville (d. 1772), a merchant of Rouen, who in 1759 presented a model to the intendant De Brou and was awarded a grant of 5,000 livres. The machine was elaborate, utilizing the labor of 148 persons, each of whom spun two threads simultaneously on a distaff before her, one with each hand. Each spinner could spin as rapidly or slowly as she wished, and could mend a broken thread without in any way disturbing the movement of other bobbins. The instrument spun not only cotton but also wool and flax.[5]

Barneville attempted to apply his invention industrially, first at Rouen and Gihors and later at Houpelines, but misfortune dogged his steps. He improved the machine and in 1766 erected one in a château near Versailles, only to see the building shortly afterwards sold and demolished. Barneville probably avoided economic ruin only through the patronage of the Countess de Brionne, who on various occasions befriended him and his family.

His son, Nicolas Denis François (1749-1842), brought the machine to a further stage of improvement. But young Barneville was one of those unfortunate individuals whose lives have been marred by war and revolution. For three years he was in the American Revolution, and during most of the French Revolution and Napoleonic era he either had a military post or was a political prisoner. His period of greatest promise as an industrialist was in the interim between the American war and the French Revolution, when he was resident in France and eager to establish an enterprise in spinning and weaving of "Indian" or painted cotton goods. He visited Switzerland in 1784 or 1785 to observe the methods by which the excellent muslins of that country were produced. In 1786 he obtained from the government, through the powerful support of Marie Adelaide and Victoire, aunts of the king, 2,400 livres and the privilege of setting up his machine in the workshop of the Hospital of the Quinze-Vingts. Later the same year he was awarded a government subsidy of 15,000 livres

---

are given in *Machines approuvées par l'Académie*, VII, 293-96. An excellent illustration of Hargreaves' machine is in Edward W. Byrn, *The Progress of Invention in the Nineteenth Century* (New York, 1900), 421.

[5] *Histoire de l'Académie, 1761*, p. 154.

a year for a period of ten years, but this proved to be little more than a grant on paper. In 1787 the government did give him 20,000 livres on condition that he furnish it with two of his machines and set them up at advantageous points in the provinces. One was placed at Rouen, only to be destroyed on July 20, 1789, by an irate mob resentful that it and some English labor-saving machines had recently been installed there. The other machine, requiring sixteen spinners, was set up in the workshop of the Hospital of the Quinze-Vingts, famous institution of the blind at Paris, where it remained in operation until 1792.

The Barneville machine produced a fine thread of excellent quality for use in muslins. One of the examiners for the Academy of Sciences reported the thread as almost twice as fine as that used in the best Indian muslins. Reportedly the machine was able to spin from a single pound of cotton 200,000 aunes (237,600 meters) of thread! This seems incredible.[6] Not even the English could equal it; the thread produced by Crompton's mule was coarser. It was not an easy machine to operate; nevertheless, a woman demonstrator in Rouen in 1789 spun with it 194 aunes (252 yards) of thread in an hour.

Interest in the Barneville machine was revived by the National Convention, which on 7 frimaire an III (November 27, 1794) passed a decree advancing the inventor 200,000 livres for ten years without interest, and shortly later one of its committees granted him use of a former convent on Rue du Bac. Barneville was immediately given 120,000 livres, but in rapidly depreciating assignats; the rest was paid him some months later on his renewed request.[7] He was discouraged rather than elated, however,

[6] Ballot, *Machinisme*, 62. This was in 1816, but the machine had not been improved since the 1780's. *Nouvelle biographie générale*, VII, 457, says 300,000 aunes a pound; André Rémond, *John Holker, manufacturier et grand fonctionnaire en France au XVIIIe siècle, 1719-1786* (Paris, 1946), 65-66 n. 155, says 300 meters a gram, or 136,265 meters a pound. These last two writers give their figures for the 1700's.

For the story of the two Barnevilles, see Ballot, *Machinisme*, 57-63; *Nouvelle biographie générale*, VII, 458-59; Tuetey, *Répertoire générale*, III, xiii, 539-40; Charles Schmidt, "Les débuts de l'industrie cotonnière en France, 1706-1806," in *Revue d'histoire économique et sociale*, VII (1914-1919), 29-30.

[7] Ballot, *Machinisme*, 61-62; *Nouvelle biographie générale*, VII, 459. On paying him the residue of this sum, the government suppressed his pension of 2,000 livres.

at the turn of events, and it appears that little came of the matter, for in December, 1798, he returned to his military post.

With the Restoration Barneville appealed once more for aid, but the government, in his opinion, was too niggardly in the support offered. Since the machine was a costly one and he did not have the private funds to promote it (his father before him had spent 80,000 livres on experimentation), he decided to renounce the matter and live on the military pension to which his long years of service entitled him. He died in 1842 at the age of ninety-three. It seems amazing that a family with so much influential backing as he and his father enjoyed in having several members of the royal family and of the nobility as patrons did not receive better government support than it did. Young Barneville was clearly a victim of the troubled times. Probably no French textile machine of the 1700's had more real merit, and yet it was a failure at a time when France badly needed machines of its type to quicken her Industrial Revolution.

A third machine for the multiple spinning of cotton was invented in 1784 by Lhomond, after the famous "spinning jenny" of Hargreaves was known. It had several useful features, permitting the operator to work seated, stopping automatically when a spool was filled, and producing a soft, elastic thread. Approved by commissioners of the Academy of Sciences, in 1787 it brought Lhomond an award of 780 livres and a pension of 2,400 livres a year on condition that he would install it in places designated by the government. Complaints, however, came from its operation at Couvanges and Bar-le-Duc, where it was first introduced, and the government January 1, 1789, stopped Lhomond's pension. On Lhomond's protest, a new examination was given the instrument in September, 1791, when it was placed in competition with the machine used by the Englishman Milne, whose work in France the government was then subsidizing. In the opinion of the judges, Lhomond's machine was the better. It was criticized, however, for not producing in larger quantity and for not having means of carding materials. The next year Lhomond invented a carding machine and received a grant of 6,000 livres, the highest sum allowed to inventors at that time.[8]

[8] Ballot, *Machinisme*, 51-52; Young, *Travels in France*, 96. Young called Lhomond "a very ingenious and inventive mechanic."

Other inventions for spinning were claimed,[9] the only one of interest being that of Frédéric Hildebrand, a naturalized Swiss who had come first to Lyons in 1753 and later to Paris, and made a number of alleged inventions. Among them was a machine for the spinning of thread by the blind, employing twelve workers. Prior to the Revolution he had introduced a machine of this type for the blind children at the house of the Philanthropic Society in Paris and devoted time gratuitously to their instruction in its use.[10]

As for silk, spinning is not necessary inasmuch as it comes from the cocoon in a thread. But it must be unwound and thrown (or doubled), and during the 1700's several processes for drawing and winding it were invented.[11] Among them were machines by Villard (1767), Tabarin (1760's), the Abbé Decène and Audemar jointly (1769), Rival (1783), Etienne Suchet (1783), Bouceret (1789), and Paul Paulet (1792). That of Villard, begun in 1760 at Lyons, by eliminating one or more steps in the recognized process speeded it up. It was ridiculed by Vaucanson, but Villard improved it, applied it industrially in 1767, and in 1779 took it to Paris and gave demonstrations at the Tuileries during 1779-1780 before commissioners of the Academy of Sciences, who were pleased with its performances and made a highly commendatory report. The government in 1782 gave Villard a grant of 5,000 livres and a pension of 3,000 livres, paid until his death in 1789. His machine, however, was only used in manufacture at Salon, near Lyons.

Tabarin's instrument was an improvement of one which Vaucanson had in turn bettered earlier in the century. It was developed and first applied in the 1760's in Languedoc, where officials twice gave him a reward of 1,000 livres. Its use spread in the 1770's and 1780's along the Rhône valley and in Provence.

---

[9] See Tuetey, *Répertoire générale*, III, 542; Ballot, *Machinisme*, 230; Ballot, "Girard," 137-38.

[10] Hildebrand also invented a machine for cleaning and carding hemp and flax and an instrument for overcoming deafness. He never received a government grant or pension, but he was a protégé of Bertin, intendant of Lyons and later of Paris. Archives nationales, F15, 3596.

[11] Lacroix, *XVIIIme siècle*, 532; Ballot, *Machinisme*, 328 n. 2, 329-30; Tuetey, *Répertoire générale*, III, 539; *ibid.*, VII, 269; *Inventaire-sommaire des archives départementales antérieures à 1790. Bouches-du-Rhône*, ed. by Louis Blancard (Paris, 1865), C 88 (hereafter, *Archives des Bouches-du-Rhône*); *Archives parlementaires*, XIV, 318.

In 1796 it was given national recognition, and it enjoyed its period of greatest use during the Napoleonic era.

The invention by Rival attempted to make the drawing of silk automatic. He demonstrated some of his machines near Lyons in 1783. The government manifested interest in them and subsidized their installation in certain factories. Rival claimed that his machines were a fourth faster than others in use at that time, and his product better.

The machine by Bouceret was pronounced by commissioners of the Academy of Sciences in their report (*ca.* 1791) preferable to those of Vaucanson and Villard, but other information on it is lacking. Evidently a machine by Paulet for drawing silk was not a bad one. For it and other inventions he was awarded in 1792 the unusual sum of 10,000 livres by the Revolutionary government.

In the field of weaving, certain processes were invented by Philippe de Lasalle (1723-1804), a native of Seyssel, who was trained as a painter under Sarrabat and Boucher. Setting out for Rome to pursue his studies further, he met en route a Lyonnaise merchant who offered him a business partnership and his daughter's hand in marriage. Thenceforth Lasalle resided at Lyons and gave his attention to the development of processes in the manufacture of elegant cloth. His first and perhaps greatest achievement was an improvement in the loom used in this manufacture (about 1775), so that work which once had required at least two months could be done in a few minutes.[12] This was ten years before England had Cartwright's loom, which even then needed improvements before it was of much service. For this achievement Turgot obtained for Lasalle a pension of 6,000 livres and the decoration of Saint-Michel. Later, in 1783, for further improvement of his loom, he was given the Grand Gold Medal for achievements most useful to commerce. He is reported to have invented during Necker's first ministry (1776-1781) a ma-

---

[12] A discussion of this process is given in the Paris *Moniteur universel*, April 2, 1804, and summarized in a note following the sketch of Lasalle in *Nouvelle biographie générale*, XXIX, 726. See also Pierre Foncin, *Essai sur le ministère de Turgot* (Paris, 1877), 290-91; *Archives parlementaires*, XIII, 470; Turgot, *Oeuvres*, IV, 643-44.

THE LOOM OF JACQUES VAUCANSON

This machine, invented in 1746, wove patterns in silk.

(Courtesy of the Conservatoire National des Arts et Métiers, Paris, where it rests)

chine with flying shuttle for the making of gauzes, veils, and other materials, in any length desired.

Another inventor whom Turgot desired to aid was Leturc (b. 1748), a native of Lille who served for a time as engineer under the Administration of Bridges and Highways and later as professor of the Royal Military School at Paris. While in this latter position he invented an elaborate loom for the manufacture of lace, which was examined for the Academy of Sciences in 1776 by Vandermonde, Vaucanson, and Laplace. Their report was eulogistic. Leturc's machine was the first for the manufacture of lace. Nevertheless, due apparently to Turgot's fall from power, Leturc failed to receive the expected financial support from the French government. Piqued at his country's ingratitude and unable to get legal redress against a worker who stole his invention, he broke his machine and went to England, where for ten years he lived and trained a large number of English technicians in lace manufacture.[13]

The Rouen inventor Biard likewise experienced difficulty in gaining recognition of a loom said to have been invented about 1795, but not given a patent until 1804. In 1801 he entered into partnership with a certain Dodard for use of ·the invention, and set up thirty looms in their workshop at Rouen. Shortly later Napoleon, while passing through the city, visited the factory and manifested much interest in the machine. He gave Biard an award of 1,200 francs. This invention has been designated by Charles Ballot as "the first mechanical loom which functioned efficiently in France." The "entirely automatic" loom was powered not by water or steam but by hand, and a piece of cloth two aunes (2.59 yards) wide by two aunes long represented a day's work. This was far from rapid.[14]

Looms for the weaving of silk were invented by Jurine (1717), Vaucanson (the 1740's), Jaubert and Rocamus of Provence (1767), Rivey of Rivet (1790), and Jean Paulet (1792).[15] Vaucanson de-

[13] Ballot, *Machinisme*, 273-74.
[14] *Ibid.*, 252-53. Each machine cost from 1,500 to 2,000 livres, and a month's output of goods was valued at 180 livres.

[15] J. B. Monfalcon, *Histoire monumentale de la ville de Lyon* (9 vols., Paris and Lyon, 1866-1869), III, 3-5; Tuetey, *Répertoire générale*, III, 539; *ibid.*, VII, 262; *Collection de documents*

vised several looms for weaving intricate designs. One machine bought for the city of Tours cost 100,000 livres. Textile workers feared Vaucanson's machine would end their employment, and the hated inventor was threatened and stoned. It is said that in revenge he designed a machine "by means of which an ass wove a piece of flowered silk."[16] Jurine, a Lyonnais lacemaker, is reported to have invented a good loom in 1717. In 1742 an unnamed inventor is credited with "having made a machine truly marvelous," by which cloth of gold could be woven without seam.[17] To Claude Rivey, a Parisian technician, the government early in 1792 gave 6,000 livres, the maximum award at that time, for a loom which would weave either silk or cotton.[18]

It was by a series of notable inventions in silk weaving that the way was paved for the remarkable invention in 1801 of the Jacquard loom. This masterpiece stands even today as a crowning work of industrial art, by which an elaborate design in colored silk can be mechanically and elegantly woven.

Certain inventions allied to weaving also played a notable role in French textile progress. These were inventions in the field of painted cloths. Around 1756 Bonvallet, a manufacturer of Amiens, had invented a machine for printing cloth by means of an engraved plank fitted into a cylinder equipped with an iron wheel. Heat was supplied by iron bars previously heated being inserted into the wheel. The impression was made slowly, but it was good, in fact clearer than that formerly made by an engraved board alone. The invention appears to have been utilized only at Amiens, and there by only one industrialist other than Bonvallet.[19]

---

pour servir à l'histoire des hôpitaux de Paris, ed. by Michel Möring and Charles Quentin (4 vols., Paris, 1881-1887), II, 273 (hereafter, *Documents des hôpitaux de Paris*); *Archives des Bouches-du-Rhône*, C 85; Lacroix, *XVIIIme siècle*, 532.

[16] This machine was invented in 1744, at the time of an economic depression. Monfalcon, *Histoire de Lyon*, III, 3; François Dumas, *La généralité de Tours au XVIIIe siècle: administration de l'intendant Du Cluzel (1766-1783)* (Tours, 1894), 149-51.

[17] Lacroix, *XVIIIme siècle*, 532, citing the *Journal de Barbier*.

[18] Tuetey, *Répertoire générale*, III, 539; *ibid.*, VII, 262. Apparently this was the same man as Rivet the carpenter, mentioned in *Documents des hôpitaux de Paris*, II, 273, and the silk loom the same, despite the date of 1791 given for the invention and the government award.

[19] Ballot, *Machinisme*, 290. The dye was applied by hand.

A second machine, by an unknown inventor, was in use when described in 1780 by Roland. It was composed of three wooden cylinders, one of which carried a bronze impression. Beneath it were two cylinders the lower of which dipped into a tub of dye and carried the color to the upper cylinder. The cloth to be dyed passed between the top two cylinders, and after the impression was imparted to it went around a series of small wooden rollers. The whole was controlled by a handwheel.[20]

In 1791 a new machine for printing cloth and paper in large quantities was invented by Robillard, a Parisian machinist. After due investigation the Bureau of Consultation of the Arts and Trades recommended that the government award him 4,000 livres, partly by way of reimbursement for heavy expenses and partly to encourage him to perfect certain other machines with which he was toying.[21]

The most notable invention of this sort during the century, however, was that by Oberkampf and his nephew Samuel Widmer in 1797. It was made and applied at Jouy, where Oberkampf's manufacturing establishment was situated. Oberkampf was the son of a German industrialist who after some years of manufacturing in southern Germany moved with his family to Switzerland and set up an establishment to produce printed goods (called *Indiennes*). The son after receiving such training as his father was capable of giving apprenticed himself as an engraver to a royal manufacturer in Lorraine. Then he went to Paris to accept a position with Cottin, a cloth printer on the Bièvre River. Shortly he was offered a position and then a partnership by another manufacturer, which he accepted on condition that their factory be set up at Jouy-en-Josas, near Versailles along the Bièvre, whose waters were erroneously thought good for dyeing. There, after some years of slow development, Oberkampf built a workshop renowned throughout France for the high quality and beauty of its output. He imported directly from the Orient much of his raw material and finished it in his shops. Around a thousand workmen were employed at Jouy alone, not to speak of one or more branches directed by his brother Fred-

[20] *Ibid.*, 291.   [21] Tuetey, *Répertoire générale*, VII, 262-63.

erick. On the eve of the Revolution and during the 1790's, Oberkampf's holdings, valued in the millions, made him one of the wealthy men of France. Proximity to the court at Versailles no doubt had much to do with his fame and prosperity. It might hastily be thought that Oberkampf would suffer from the Revolution; on the contrary, he catered to the Revolutionists through some large gifts to patriotic enterprises and in consequence was never suspect. His wealth and renown he continued to enjoy under Napoleon, who with Josephine paid a visit to Jouy; on their departure the emperor removed his own ribbon of the Legion of Honor and with it decorated Oberkampf.

While Oberkampf promoted it, used it, and possibly even suggested it, the actual work on the invention was done by his nephew, Samuel Widmer, whom he had reared and trained as a chemist, engaging Berthollet to give him instruction. The invention consisted of a copper cylinder on which was engraved in depression the design to be printed. Pivoted with metal instead of wood, it was operated by mechanical power at a faster rate than earlier types. It was used first in September, 1797, and easily printed 5,000 meters of cloth a day. It was distinctly superior both in speed and workmanship to all the processes hitherto used.[22] Widmer improved the cylinders in 1801 and subsequent years, and in 1806 he was awarded a gold medal at an industrial exposition at the Louvre.[23] With this machine it was possible in five or six days to turn out work formerly requiring six months.

Passing mention might be made of certain inventions in other fields allied to weaving. In 1792 a new and cheaper method of manufacturing collars was invented by one Grenet, for which the Bureau of Consultation of the Arts and Trades recommended that the Legislative Assembly award him 5,000 livres. It was expected that this invention would result in "opening a considerable branch of commerce and retaining in the kingdom a very large quantity of coin."[24]

---

[22] Henri Clouzot, *Histoire de la manufacture de Jouy et de la toile imprimée en France* (Paris and Brussels, 1928), 7-21.

[23] *Ibid.*, 23-27.

[24] Tuetey, *Répertoire générale*, VII, 265.

In the same year this bureau recommended to the Assembly the payment of 6,000 livres to one Anthéaume, a hat manufacturer, for an invention in the making of felt goods.[25]

Other inventions could doubtless be added. Those described, however, suffice to impress one with the fact that the French were far from inert in textile invention. France lagged behind England, however, especially in developments in cotton manufacture, and the French government from 1759 or 1760 was actively engaged in importing newly designed British machinery and British technicians. Among the large number of British textile experts induced to come were the Milnes, father and son, Foxlow, Holker, Philemon Pickford, Garnett, Morgan, Massey, and Leclerc.[26] Certain Swiss, Germans, and Irishmen likewise were recipients of government encouragement,[27] but it was England, whose recent inventions put her foremost in the textile world of cotton and wool, that France was most concerned to copy.

These foreign workers sometimes brought their machines. This was not easy, for England as well as other countries had strict laws of prohibition against the export of new machines without permission, regarding them as trade secrets.[28] Nevertheless English machines did enter France. A larger number perhaps were reproduced by the workmen after arriving in France. A government report in 1790 showed that France then had 900 spinning jennies in operation. This looked unfavorable in view of the fact that Britain at that time had 20,000 in operation, and in addition 7,000 to 8,000 of Crompton's "mule-jenny." The latter machine appears to have been used first in France in the velvet factory of Morgan and Massey at Amiens in 1789.[29] The Watt and Boulton steam engine, too, was imported into France for service on the eve of the Revolution.[30] Unfortunately an economic depression developed in France after 1787 and reached such severity that unemployment and famine engulfed the coun-

[25] *Ibid.*, 263.
[26] Numerous references to these men are *ibid.*, vols. III and VII, and in Ballot, *Machinisme*. See also Schmidt, "Industrie cotonnière," 26-29; Clouzot, *Manufacture de Jouy*, 152-53.
[27] See Clouzot, *Manufacture de Jouy;* Tuetey, *Répertoire générale*, III, 541-42.
[28] Schmidt, "Industrie cotonnière," 27-28.
[29] *Ibid.*, 27.
[30] Ballot, *Machinisme*, 396.

try and lasted until 1791. The problem was to create jobs, not to eliminate them, as the new machinery was charged with doing. There were riots against it in various cities, and it was deemed prudent to discontinue use of some of the English machines not already destroyed. All in all, the outbreak of the Revolution in 1789 hampered industrial enterprise; nevertheless it is a curious fact that perhaps no period of the eighteenth century saw the birth of so many inventions, textile and otherwise.

It is customary to date the beginning of the Industrial Revolution in France from the Restoration era. Certainly the steam engine and the new large-scale manufacturing machines made only faint development prior to that period. Another feature of the Industrial Revolution, the factory system, was more advanced. There were scores, possibly hundreds, of manufacturing establishments, above all in the textile line, which employed large groups of workers ranging from fifty to a thousand, working with tools and equipment provided by the capitalists.[31] The Oberkampf works at Jouy employed one thousand workers. Another of the larger establishments was that of Decrétot at Louviers, an important manufacturing town. This establishment occupied a large two-floored building with a magnificent front, the whole costing 209,906 livres, and its product of woolen cloth was praised by Arthur Young as the most elegant in the world.[32] There were also dozens of municipal asylums, known as general hospitals, which had sewing rooms and factories employing several score workers. Some gave employment to dayworkers as well as to inmates. So far as the factory system was concerned, France was rapidly entering on it well before 1789.

---

[31] See, for instance, Clouzot, *Manufacture de Jouy*, 68-172. Most of the figures here are for 1806, when a government census was made, and there were hundreds of factories with several score or more workers. The picture for the 1700's is given *ibid.*, 78, 86, 87, 99, 107, 119, 129, 130, 143, 158, 159, 160, 165, 172.

On large-scale employment in the glass industry, see Scoville, "State Policy," 431.

[32] "I had letters for the celebrated manufacturer Mons. Decretot, who received me with a kindness that ought to have some better epithet than polite; he shewed me his fabric, unquestionably the first woolen one in the world, if success, beauty of fabric, and an inexhaustible invention to supply with taste all the cravings of fancy, can give the merit of such superiority. Perfection goes no further than the Vigonia cloths of Mons. Decretot, at 110 liv. (41. 16s. 3d.) the aulne." Young, *Travels in France*, 144.

Chapter VIII

Automata

Vaucanson (1709-1782), the celebrated inventor from Grenoble, was the creator of several automatic figures, such as a flute player, a tambourine player, a duck, and an asp. So lifelike in action were all these that they attracted enormous public attention. The flute player came first in point of time. It had a repertoire of twelve tunes, displaying a great range in notes in three octaves. The idea of this robot was suggested to young Vaucanson when he first came to Paris and saw the statue of a flute player in the garden of the Tuileries. He spent several years at the task, as he had to learn more about the several sciences involved. An uncle learned of his scheme and was so provoked at its apparent foolishness as to consider asking for his imprisonment on a *lettre de cachet*. Vaucanson decided to leave the capital for a while, and accordingly went to Normandy and Brittany, where he spent three years in roaming and study. Finally returning to Paris, he completed his invention in 1738. Condorcet relates that when the final moment came for a test, Vaucanson sent his servant out on an errand but that the servant, suspecting his master's intention, hid in the room. When the notes poured forth from the instrument with a naturalness that would equal the fondest hopes, the servant bounded forth from his hiding, fell on his knees, and embraced his master with tears of joy. The public skeptically heard of the invention. In general it was regarded as a piece of quackery until the Academy of Sciences, to whom it had been submitted, pronounced it a genuine, remarkable invention, imitating perfectly both the movements and the notes of a human flutist. Cynicism gave way to admiration, and overnight Vaucanson emerged as a prodigy. The automaton was five and a half feet high, "a real flutist, playing on a real flute with his mouth and moving his lips and rendering his variations with the utmost accuracy by the aid of his fingers." Each note was produced by the tongue, and the tone was perfect. Vaucanson relates how he overcame the difficulties in achieving this.[1]

Shortly afterward Vaucanson invented an automatic shepherd which played simultaneously a flageolet *(galoubet)* and a tambourine. It had a repertoire of twenty pieces for minuets and counterdances, and throughout one of them the shepherd whistled in accompaniment. Clever though this robot was, it did not attract the popular attention that the flutist did.

Vaucanson's second most popular automaton was his duck, which waddled, quacked, flapped its wings, drank, and ate in a likelife manner. It thrust forward its head and ate food which later it defecated in a sort of semidigested form.

Vaucanson invented a mechanical asp in 1741 for use in Marmontel's play, *Cléopâtre*. The automaton would hiss and throw itself forward at the heroine's bosom. One wag of the day commented that he would like to be the asp.[2]

Vaucanson also considered creating a robot to illustrate the circulation of the blood. He was gratified by some experiments, but he decided that he must go to Guiana to obtain the "elastic gum" (rubber) necessary for forming the veins and arteries and there do the work. He applied to the government for permission for the trip, but the vexing slowness of its procedure led him to give up the undertaking.[3]

Vaucanson presented his automatic creations to the queen, who reportedly did not value them highly; yet they were placed on exhibit in the Jardin du Roi for the crowds of spectators that came to watch them. According to one account they were later dis-

---

[1] Jacques de Vaucanson, *Le mécanisme du fluteur automate, presenté à messieurs de l'Académie royale des Sciences* . . . (Paris, 1738), 3-9, 21; Condorcet, *Oeuvres complètes*, II, 416-19; Louis Ducros, *French Society in the Eighteenth Century*, tr. by W. de Geijer (London and New York, 1927), 114.

[2] Vaucanson, *Fluteur automate*, 10-17, 19-22; Condorcet, *Oeuvres complètes*, II, 420; Ducros, *French Society*, 114; *Nouvelle biographie générale*, XLV, 1019-20. Vaucanson insisted that the mechanism of each of the automata described in his book (the fluter, tambourine player, and duck) was completely different from the others and an independent invention. He also insisted that they did not rest upon the mechanism of medieval cathedral automata.

[3] Condorcet, *Oeuvres complètes*, II, 429-30.

persed, the flutist and the tambourine player going to Germany. The reputation of these robots brought Vaucanson an invitation from Frederick the Great to come to Prussia, but according to Condorcet he declined the offer on patriotic grounds, preferring to live and work in France. He remained silent about the offer, except to let Cardinal Fleury, the king's chief minister, know of it. Shortly afterward he was appointed by the French government inspector of silk manufactures in the kingdom, a post that led to the development of his interest in the silk-textile field and to some remarkable inventions in it.[4]

Vaucanson was by no means the originator of automata, which indeed had their origin in ancient times. Archytas, a friend of Plato, allegedly was the inventor of the automatic pigeon, and the Middle Ages gave birth to a steadily increasing number of robots, most of them in connection with cathedral clocks.[5]

Besides Vaucanson, several other gifted Frenchmen of the 1700's gave their attention to the invention of automata. Either in the late 1600's or early 1700's two ingenious mechanical tables were constructed by Abbé Truchet (1657-1729), a skilled mechanician of the Carmelite order who assumed the name of Père Sebastian. One of these tables, called by Louis XIV his opera, was slightly more than sixteen inches long and an inch in thickness, on which a multitude of tiny creatures emerged from hiding at a given signal and danced or gestured. Five different acts were presented on this tableau. The second table, larger and yet more intricate, displayed at a given signal an animated landscape. "A river ran; some tritons, some sirens, some dolphins swam from time to time in a sea that bordered the horizon; one hunted, one fished; some soldiers went to mount guard in a citadel stationed on a mountain; some ships came into port; [and] Père Sebastian himself was there departing from a church en route to thank the king for a grace newly obtained, for the king passed by hunting

---

[4] *Nouvelle biographie générale*, XLV, 1020; Ducros, *French Society*, 114.
[5] *Cyclopaedia of the Industry of All Nations*, ed. by Charles Knight and George Dodd (New York and London, 1851), 268; Alfred Ungerer, *Les horloges astronomiques et monumentales les plus remarquables, de l'antiquité jusqu'à nos jours* (Strasbourg, 1931), 16. The latter work has numerous illustrations of medieval automata.

with his group." Such is the description given by Fontenelle in his éloge of Abbé Truchet.[6]

Other robots were made by the Lorraine physicist François Joseph Camus (1662-1732), also a member of the Academy of Sciences at Paris, who in the late years of Louis XIV's reign worked at that monarch's request to construct a company of soldiers which would march and defile for the amusement of the dauphin. The mechanical toy was not completed when the king died in 1775, and Camus discontinued work on it. Later in life when he had acquired a national reputation as a physicist and had become a member of the Royal Academy of Sciences, he made a small automatic carriage with drivers and passengers, fashioned to move by a clocklike mechanism.[7]

Another early eighteenth-century designer of automata was one Maillard, who in 1733, five years before the invention of Vaucanson's flute player, submitted to the Academy of Sciences detailed plans for a carriage drawn by a mechanical horse and also for a mechanical swan which would swim on a pond. Both devices received the approval of the academy (equivalent in that day to a patent).[8] There is no evidence, however, that Maillard ever constructed the mechanical swan and the mechanical horse. Whether they would have been successful is thus not altogether clear. The drawing of the mechanical horse appears somewhat fantastic; nevertheless the academy was convinced that it would work.

Interestingly enough, Vaucanson's claim to the invention of the mechanical duck was challenged in court by Dominique François Bourgeois de Châteaublanc (1698-1781), a physicist of no mean ability. Bourgeois lost the contest and for two and a half years was confined to a Paris prison as a calumniator. Later he proved his skill by the invention of a street lamp which won a prize offered in 1764 by the Academy of Sciences, and by the invention in 1769 of a lighthouse beacon visible seven leagues (seventeen and a half miles) and safe against winds and storms.

[6] Bernard Le Bovier de Fontenelle, *Oeuvres de Monsieur de Fontenelle* (new ed., 10 vols., Paris, 1758), VI, 395-98.

[7] *Nouvelle biographie générale*, VII, 425-27.

[8] Minutely described with drawings in *Machines approuvées par l'Académie*, VI, 133-35, 141-45.

Some years later (1778) he went to Russia on the request of Catherine the Great and constructed a beacon *(fanal)* for the port of St. Petersburg. One contemporary writer persisted in the opinion that he was the creator of the mechanical duck.⁹

Another creator of automata, the Abbé Mical (1730-1789 or 1790), fashioned a group of mechanical flute players which were praised by the writer Rivarol both for their beauty and for their musical perfection. Unfortunately criticism from certain quarters was directed at these players because of their nudity, and the Abbé destroyed them. He then undertook to construct some talking heads. The first, made of brass and uttering some short phrases, he demolished after an eulogistic article about it appeared in a newspaper. A strange personality this Abbé Mical, who winced both under criticism and praise! Inconsistently enough, he now set about the construction of two more talking heads and presented them in 1783 to the Academy of Sciences. Each was set on a box within which was a mechanism for producing artificial speech. Their imitation of the human voice was not perfect, but the inventor was highly commended on his creation. According to Rivarol, the phonetics were produced by two cylinders, each giving a certain number of words and phrases in an ingenious manner. The French government, on the advice of Lenoir, lieutenant of police of Paris, refused to buy the two heads. Reports vary on what happened to them, and indeed what happened to the Abbé. According to one source, he sold the heads to an undesignated purchaser; according to another, he destroyed them in a moment of despair. He died in poverty.¹⁰

About 1782 the Duc d'Orléans constructed an automatic cannon at the Palais-Royal, in the heart of Paris, to fire each day at noon. The powder was ignited by a fixed burning glass *(verre ardent)*. The idea of the cannon was suggested to him in 1777 by the Comte d'Angiviller, director of royal buildings, who proposed using it at the Samaritaine (an old royal building of note carrying on its front a representation of the Samaritan woman), to replace an old undependable clock and carillon. This cannon at the

---

⁹ *Nouvelle biographie générale*, VII, 78-79. The author alluded to is P. Joly, *Mémoires sur les lanternes à réverbère* (Paris, 1764).

¹⁰ *Nouvelle biographie générale*, XXXV, 312-13; *Biographie universelle*, XXVIII, 517-19; Lacroix, *XVIIIme siècle*, 70.

Palais-Royal made a considerable impression upon those who saw it.

About the same time an automatic gong, invented by Buffon and the architect Verniquet, was set up in the Jardin des Plantes. At the highest point in the garden an iron kiosk was erected, surmounted by an armillary sphere with a globe at its center representing the earth. Attached to the sphere was a hammer suspended in the air by a horsehair. At noon each day the hair was burned in two by a lens, and the hammer fell against a gong.[11]

Interest in automata was displayed at this time by inventors in other countries. At Chaux-de-Fonds, Switzerland, Pierre Jacquet Droz (1721-1790), an ingenious mechanic who showed his skill in many ways, invented an automatic writer which imitated perfectly the movements of the human hand in writing. His son, Henri Louis Jacquet Droz (1752-1791), educated in part at Nancy and no less adept at mechanics than his father, invented among other things an automatic girl harpist who followed sheet music with her eyes and at the conclusion of the piece arose and saluted her audience.[12] At Vienna, Wolfgang von Kempelen (1734-1794), a native of Pressburg, invented a remarkable chess player (1783) and a talking machine.[13]

Certain ingenious automatic devices illustrating movements of the heavenly bodies were invented by eighteenth-century French astronomers. Réginald Outhier (1694-1774), a Burgundian priest, drew plans in 1726 for a celestial sphere, five inches in diameter, which would illustrate all the phases of the moon. His compatriot, J. B. Caton, constructed the machine, and it won the praise of scientists and the election of Outhier as corresponding member of the Academy of Sciences. He later accompanied Maupertuis in 1736-1737 on his expedition to Lapland to measure a degree of longitude, and published an account of it in 1744.[14]

[11] Alfred Franklin, *La vie privée d'autrefois: arts et métiers, modes, moeurs, usages des Parisiens du XII⁰ au XVIII⁰ siècle d'après des documents originaux ou inédits* (1st ser., 23 vols., Paris, 1887-1901), IV, 136-38.

[12] *Nouvelle biographie générale*, XL, 812-13. Lacroix, *XVIIIme siècle*, 70, gives an illustration of the automatic writer. See also Timbs, *Wonderful Inventions*, 124-25, for the account of another set of automata by one of the Drozes.

[13] *Nouvelle biographie générale*, XXVII, 540; Lacroix, *XVIIIme siècle*, 70.

[14] *Histoire de l'Académie*, 1727, p. 143; *Nouvelle biographie générale*, XXXVIII, 982-83.

THE PLANISPHERE OF ABBE OUTHIER

This mechanism, built in 1731, illustrates by its movements the phases of the moon.

(Courtesy of the Conservatoire National des Arts et Métiers, Paris, where it rests)

A second device of this sort was constructed in 1749 by Claude Simon Passement (1702-1760), an astronomer and optician. This planetarium likewise was designed to illustrate the movements of the earth, sun, moon, and planets. It was regarded as a brilliant achievement, and in recognition Passement was given a pension of 1,000 livres and a room at the Louvre.[15] A still larger and more elaborate planetarium was afterward invented by the Jesuit Lot for teaching astronomy to the young Comte de Levis. It was a sphere eight feet in diameter and could be made to move at will. Its chief function was to represent the movements of the various planets around the sun.[16]

Among the automatic devices of the century must be mentioned the Café Mécanique, an invention of the French anticipating the modern Horn and Hardart "Automats" of New York City, where one deposits nickels in a slot, turns a lever, and opens a door to the article of food desired. The Café was in existence in 1786 at the corner of Rue Montpensier and Rue des Bons-Enfants. Its proprietor, one Belleville, sought a concession for it at the Palais Royal, focal point of entertainment in pre-Revolutionary Paris, but he was unable to obtain this highly advantageous and expensive location and had to content himself with his quiet, secluded nook. By 1788 he had gone to Bordeaux to direct an entertainment enterprise, and had left in charge or sold out to an associate named Tantès (or Taréz, according to one account). Tantès succeeded in making the Café one of the curiosities of Paris. Little frequented during the summer, it was crowded in winter by numbers of both sexes and all classes, who flocked there for their evening diversion. Unfortunately the room was small and could not accommodate all who came. In common with other cafés of the period it carried a stock of current newspapers, and mention is made that one had to arrive early in the evening to be certain of reading them. On Sundays, on account of the crowds, the papers were locked up. In this Café were a number of marble tables, each resting on two hollow columns or supports that communicated with the cellar where the food was served. To obtain service, a customer would pull a chain on one of the

---

[15] *Nouvelle biographie générale,* XXXIX, 304-305; Lacroix, *XVIIIme siècle,* 69-70.

[16] Lacroix, *XVIIIme siècle,* 70.

table legs or cylinders, causing a bell in the basement to ring. Then a valve would open on the table and the patron would give his order, evidently by speaking tube. The order would then be filled by a dumb-waiter which rose by the column. The proprietress *(limonadière)* was at the cashier's desk on the main floor and was able at any time to communicate by a speaking tube with the servants in the cellar. The service appears to have been largely in drinks, even as it is in French cafés today.[17]

[17] *Almanach du Palais Royal utile aux voyageurs pour l'année 1786* (Paris, 1786), no. 99; *Tableau du nouveau Palais-Royal* (London, 1788), pt. 1, pp. 55-61. For this information obtained from the Bibliothèque Nationale, I am endebted to Miss Marjorie Coryn of Croyden, England, and a friend of hers in Paris. Miss Coryn has given a description of the Café Mécanique in her book, *The Marriage of Josephine* (New York, 1945), 88.

Chapter IX

Other Mechanical Devices

ALTHOUGH MUCH THOUGHT AND EFFORT WERE SPENT UPON objects of amusement like automata, a great number of more useful mechanical devices also received attention. One such device was a threshing machine. Historians credit the Scot Andrew Meikle with introducing the first successful threshing machine (1786). Before Meikle, water-turned threshers had been invented by the Scots Michael Menzies (1732) and a certain Leckie (1738), but their flails turned so violently that the grain was injured. The French mechanician Du Quet had designed one approved by the Academy of Sciences in 1722.[1] Wheat was flailed in a large rectangular bed by boards attached to a pole rotated by horse-powered ropes and pulleys. The machine apparently could have worked, but it was never tried.

Other threshing machines were approved by the Academy of Sciences in 1737 and 1762. The first, invented by Meiffren, a coast guard captain in Provence, was reported to thresh as much grain in twelve hours as could six good horses by tramping (the means then used in southern France). The later invention, by De Malassagny, operated along lines similar to the machine of Du Quet. No mention was made of its output.[2]

Seven years after Meikle's successful experiment, an automatic threshing machine recommended by the Society of Agriculture was in operation in France. It is not clear whether the unnamed inventor had heard of Meikle's invention; the *Moniteur Universel* mentions no such indebtedness. The machine threshed and win-

[1] A description and drawing is in *Machines approuvées par l'Académie*, IV, 27-29. On the Scots, see Kaempffert, *American Invention*, II, 291; William H. Doolittle, *Inventions in the Century* (Philadelphia and London, 1903), 41.

[2] *Histoire de l'Académie*, 1737, p. 108; *ibid.*, 1762, p. 193.

nowed grain completely, even returning the straw, and by it two men were able to do the work which formerly required fourteen.[3]

Several types of improved grain mills were designed during the century. One was a hand mill invented by a certain De la Gâche and approved by the Academy of Sciences in 1722. It embodied on a small scale the same principles as windmills and water mills.[4] A letter of July, 1770, mentioned a horse-drawn grain mill invented in Provence, but gives no description.[5] Shortly afterward, a hand mill operated by two men was invented by Claude François Berthelot (1718-1800), professor of mathematics at the Military School in Paris, already honored and pensioned by the government because of some inventions for military defense. This mill was placed in operation at the Hôpital de Bicêtre in 1778, and it was expected that its use would be rapidly extended and that it would bring a fortune to the inventor. But Berthelot turned his back on this prospect and renounced the monopolistic privileges granted him by the government. Later, the Revolutionaries revoked his pension, and he died forgotten and in misery in 1800.[6]

In 1786 a gardener named Biberon invented a mill for grinding wheat damaged slightly by decay. The Society of Agriculture of Laon supported his petition to the provincial government for recognition and a reward. The intendant, however, replied that schemes for using such wheat were numerous. He recognized certain deserving features of Biberon's machine and offered him tax remissions, but no grant.[7]

In 1792 two claims to invention of flour mills were made, one by Orelly, a former Benedictine of Bordeaux turned constitutional priest; the other by a certain Picard. Orelly used compressed air to power his mill, and he claimed that its use throughout France would result in a great saving to the nation. He refused any recompense for his invention.[8] No doubt the chief

---

[3] *Moniteur universel*, XVII, 626.

[4] A description and two illustrations are in *Machines approuvées par l'Académie*, IV, 37-38.

[5] *Archives des Bouches-du-Rhône*, C 1013.

[6] *Nouvelle biographie générale*, V, 703; *Biographie universelle*, LIII, 99-101.

[7] Emile Justin, *Les sociétés royales d'agriculture au XVIIIe siècle (1757-1793)* (Saint-Lô, 1935), 235-36.

[8] *Moniteur universel*, XI, 245; *Archives parlementaires*, XXXVIII, 3;

reason for the repeated inventions of flour mills was the persistent appearance of famine in France, even more frequent in the second half of the century than the first.[9]

A water-turned mill for rasping and grinding tobacco, invented by Chamoy, was approved in 1767 by the Academy of Sciences. The tobacco leaves were dropped between two cylinders carrying rapes, and the shredded bits fell into bins, from which they were removed to be powdered in small mills like those for grinding coffee.[10]

Mechanical devices receiving much attention were cranes and dredges. A ship's cargo derrick was invented by the Oratorian Père Ressin and approved by the Academy of Sciences in 1714. A large transverse pole was fixed to the hindmost mast about halfway up. To each side of this pole was attached a pulley, through which ran a long rope or cable. One end of the cable was hooked to goods to be moved; the other end ran through a basket and went over the opposite side of the ship, to be pulled when the cargo was to be lifted or released when the cargo was to be lowered. Weights could be put in the basket to lessen the force needed for lifting the cargo. A larger basket was attached to the second rope fulfilling the same functions. A fundamental factor in this mechanical device was a series of pulleys.[11]

A few years later François Joseph Camus invented a crane for digging a canal or constructing a road, and published a description of it in his *Traité des forces mouvantes, avec la description de 23 machines nouvelles de son invention*.[12] A similar crane invented by Dubois and called a "shovel for lifting loose dirt" was approved by the Academy of Sciences in 1726. It was mounted on a wheeled platform. The crane or lever was raised and lowered by turning a windlass with four handles (two iron rods extended through a crosspiece of wood). At the end of the lever was a large shovel or scoop, whose back was hinged and could

---

Tuetey, *Répertoire générale*, VII, 259, 269-70. Whether Orelly's machine measured up to his claims is not indicated. It was sent for examination to the Bureau of Consultation of the Arts and Trades.

[9] See the first two chapters of Shelby T. McCloy, *Government Assistance in Eighteenth-Century France* (Durham, N. C., 1946).

[10] *Histoire de l'Académie*, 1767, p. 184.

[11] *Machines approuvées par l'Académie*, III, 29-30.

[12] *Nouvelle biographie générale*, VIII, 425-27.

be lowered to release the dirt. The crane as shown by its drawing appears to have contained the essential features of the modern machine.[13]

Later in the century Jean Tremel (1785), Ambroise Poux-Landry and Lepoule (1790), and Desvallons (1792) improved the crane, in most of these instances used for the loading of ships. That Tremel's crane had its merits seems evident from the fact that the government gave him a pension of 500 livres. In the case of the others, there are only the claims made for them.[14] Desvallons was the inventor of two types of crane, both of them praised by the committee reporting them to the Legislative Assembly. By means of one it was stated that a single workman could do the work of six; with the other two men could do that of twenty. The second crane reputedly could lift with ease a weight of six thousand pounds.

A remarkable machine for cleaning harbors was approved by the Academy of Sciences in 1703. Designed by the Abbé Gouffé, the dredge had two huge steel jaws controlled by two ropes or cables attached to windlasses. It could be raised and lowered by a steel bar.[15] Improvements in the dredge were later made by Héricé in 1759 and Morainville in 1782, the former living in the region of Bordeaux and the latter in Marseille. Héricé pointed out in his petition for recognition that French rivers around Bordeaux were constantly filling with sand, and that boats from America had to be lightened by removal of a third or half of their cargo before they could proceed up the Garonne to Bordeaux.[16]

A nondescript motor invented by the Abbé de Mandre (c. 1728-1803) was recommended by a committee to the National Assembly in 1789. It could be used for pile driving, for aiding pumps, and for drawing barges upstream past rapids. It had successfully pulled a train of thirty boats, four of them filled with gravel, against one of the swiftest currents in the Rhine. Following the

---

[13] *Machines approuvées par l'Académie*, IV, 165-66.

[14] *Archives parlementaires*, XV, 211; Tuetey, *Répertoire générale*, III, 538; *Moniteur universel*, II, 116; *ibid.*, IX, 71-72.

[15] *Machines approuvées par l'Académie*, II, 63-64.

[16] *Inventaire-sommaire des archives départementales antérieures à 1790. Gironde*, ed. by J. B. Gras (Paris, 1864), C 3716 (hereafter, *Archives de la Gironde*); *Archives des Bouches-du-Rhône*, C 2493.

committee's report, Malouet arose to say that the machine had been tried with complete success at Toulon, where he had been intendant. The matter was referred to the Academy of Sciences, which recognized the "new and ingenious" nature of the device but insisted that its utility was not "sufficiently great to merit a large recompense." It appears, nevertheless, to have been a useful invention, but the Abbé drew nothing from it and some years later died in obscurity.[17]

Another machine invented in the latter half of the century polished iron, copper, and other metals. It was described as "very ingenious," and the government rewarded its inventor, Gabriel Le Masson, with a pension of 300 livres.[18]

The eighteenth century saw a number of improvements in hydraulic machines. There had been much experimentation in seventeenth-century Europe with pumps, by Pascal, Von Guericke, and others, and during that century had been constructed the famous "Machine of Marly" (1676-1682), which remained throughout the eighteenth century one of the wonders of engineering science in France.[19]

The monk André Ferry (1714-1773), professor in mathematics draftsmanship at Reims, constructed a horse-drawn machine that in an hour could supply the day's water needs to the Hôtel-Dieu (hospital) of Rouen in 1756-1757. Water was forced through a pipe to a large reservoir a half mile away.[20]

An improved means of drawing water by windmill was approved by the Academy of Sciences in 1767. It was invented and put in operation near Provins by a retired army officer named Dudit de Mezières.[21]

[17] *Moniteur universel*, II, 391; *Nouvelle biographie générale*, VII, 516; Paris *Mercure de France*, December 26, 1789, p. 331. Malouet was one of the most capable men in government service at the close of the Old Regime.

[18] *Archives parlementaires*, XIV, 783. This pension, first granted in 1764, was renewed in 1778 and 1788.

[19] This machine, constructed by the Belgian Rennequin Saulem, consisted of 221 pumps which forced water to the top of a hill, from where it was sent to supply Marly, Versailles, Louveciennes, and St. Cloud. The pumps were powered by fourteen giant water wheels turned by a falls in the stream at Marly. *Grande encyclopédie*, XXIII, 204; Abbott Payson Usher, *A History of Mechanical Inventions* (New York, 1929), 297.

[20] P. Théodore Legras, *Notice historique sur les deux hôpitaux et l'asile des aliénés de Rouen* (Rouen, 1827), 51; *Nouvelle biographie générale*, XVII, 566.

[21] *Histoire de l'Académie, 1767*, p. 184.

Various savants gave attention to the matter of purifying the water of the Seine for use at Paris. The academician Leroy suggested the pumping of water from the river to a great reservoir, from which it could be taken in barrels for peddling through the city. Acting on this suggestion, the Périer brothers set up at Chaillot (1781) pumps for forcing filtered water into some enormous reservoirs at a considerable elevation above the river. This plant became a curiosity to sight-seers, although it differed little, if at all, from the Machine of Marly.[22]

Charles Vincent Vera, a postal employee, designed a device for transporting water to great heights by "a cord or vertical band without end," to which apparently buckets were fastened. The government was sufficiently pleased with the device to award Vera a pension of 400 livres in 1784.[23]

Despite these various devices, Paris was not sufficiently supplied with water, and in 1787 the Academy of Sciences offered a prize for the best solution, submitted by Gondouin-Deshais. The contest evoked a number of papers. Among them was one by an engineer named Trouville (1746-1813) proposing the transportation of water from the Seine to all parts of the city by means of a siphon. This scheme won honorable mention from the National Assembly, whose committee of commerce and agriculture praised it highly, hailing it as one which would supply all cities with water and even irrigate agricultural lands. The Assembly, however, called for more careful scrutiny of the project by an enlarged committee, and apparently it did not pass muster, since it is not mentioned again in the records.[24]

Other hydraulic devices were proposed by Reynard, a mechanic, in 1791, and by the Montgolfier brothers and Argand in 1796. Both were for supplying buildings with water. Of the former no description is given save that it was considered an im-

---

[22] Lacroix, *XVIIIme siècle*, 66. The reservoirs were 166 feet above the Seine.

[23] *Archives parlementaires*, XV, 225. On the other hand, this might be the device illustrated in *Encyclopédie méthodique*, XIX, pl. 49. Two coiled pipes, the lower ends of which are fastened to a large wooden float just below the surface of the water, are connected by ratchets at the top to a water wheel. As the pipes are rotated, they pick up water at the lower end and eject it from the upper end into a container. The *Encyclopédie* does not name the inventor.

[24] *Archives parlementaires*, XXII, 734; Tuetey, *Répertoire générale*, III, 538; *Biographie universelle*, XL, 597-98.

provement over those then in use in Paris.[25] That by the Montgolfiers and Argand used an automatic valve to control the flow of water through pipes from a reservoir to a building. In 1772 the Englishman John Whitehurst had invented a "ram" for carrying water from a reservoir to a building through a series of pipes, but his stopcock or valve, which had to be opened or closed by hand, left much to be desired. For this stopcock the Montgolfiers and Argand in 1796 substituted "a loose impulse valve in the waste pipe, whereby the valve was raised by the rush of the water, made to set itself, and check the outflow and turn the current into the air chamber." "This simple alteration," it is said, "changed the character of the machine entirely, rendered it automatic in action and converted it into a highly successful water-raising machine." Where formerly water could be raised four to six feet, the new automatic valve enabled it to be raised thirty feet. In consequence, water could be drawn in larger quantities than needed, and the public found it so useful that it was widely employed throughout the nineteenth century. In recognition of the merit of this new device, the French government at its exposition of 1802 awarded Montgolfier a gold medal.[26]

Among the intriguing mechanical inventions of the period was an instrument for measuring the distance covered by a vessel. Invented by Pourcheff, it was approved by the Academy of Sciences in 1719. It was placed on the side of a ship and consisted of a belt which passed over three wheels in triangular arrangement. One of the wheels was attached to a stem with a ratchet which turned a fourth wheel, after the fashion of a clock. Inside the vessel was a clocklike dial on which were three needles pointing to numerals on several concentric discs. By observing the positions of the needles on this dial, the pilot was able to calculate the distance covered in leagues.[27]

Two odometers were invented in the earlier half of the century: one, by Meynier, was approved by the Academy of Sciences in 1724, and the second, by Hillerin de Boistissandeau, in 1744. Both were attached to the wheel of a carriage and recorded

[25] Tuetey, *Répertoire générale*, III, 538-39.
[26] *Mémoires de l'Institut*, II, 127; Doolittle, *Inventions*, 169; Etienne Pacoret, *Le machinisme universel*, ancien, moderne et contemporain... (Paris, 1925), 145.
[27] *Machines approuvées par l'Académie*, IV, 203-204.

the number of revolutions. Even the first was described as accurate and hailed as a valuable aid to cartography; the second was regarded as more accurate and ingenious.[28]

A French engineer named Alexis Jean Pierre Paucton (1736?-1798) suggested in 1768 the use of screw propellers for ships, but not until several decades later was the idea successfully applied by two Englishmen, Smith and Rennie. Paucton later became a member of the Institute of France. It appears, however, that Daniel Bernoulli in 1752 proposed the idea, and so did Watt in 1770. In 1794 Lyttleton in England took out a patent for one.[29]

A delicate torsion balance for the measurement of electrical repulsion, invented in the early 1780's by Charles Augustin Coulomb (1736-1806), was one of the more brilliant eighteenth-century French inventions, and its usefulness and wide adaptability have given it permanent value. The central feature of this instrument was a vertical silver wire, or any other elastic fiber, from which was suspended a slender horizontal crossbar—a piece of straw covered with sealing wax, to one end of which was attached a pith ball and to the other a small circular piece of paper as counterweight. To prevent air disturbances, the whole was enclosed in two glass cylinders, one superimposed on the other. The upper cylinder, enclosing the wire, was of small diameter; the lower, enclosing the horizontal straw, was necessarily much larger. At the top of the lower cylinder was a small opening through which a disturbing force caused by a similarly charged body could be brought into close proximity to the electrically charged pith ball, and on the circumference of this cylinder was a scale in degrees by which the torque or distance moved by the pith ball under repulsion could be determined. Force was then applied to the torsion micrometer at the top of the smaller cylinder to which the wire was suspended in order to restore (by twisting) the crossbar to its original position. By measuring this torque, it was possible to know the force of repulsion exerted by the second of the two bodies under consideration. Modified forms

---

[28] *Ibid.*, 101-103; *Histoire de l'Académie*, 1744, p. 61. Mention is made of a pedometer, which apparently suggested the odometer.

[29] Louis Figuier, *Les grandes inventions scientifiques et industrielles chez les anciens et les modernes* (Paris, 1859), 139-40; Timbs, *Wonderful Inventions*, 268; Doolittle, *Inventions*, 442.

THE TORSION BALANCE OF CHARLES COULOMB

This particular instrument, built in 1898, is a reproduction of Coulomb's device for measuring electrical repulsion.

(Courtesy of the Conservatoire National des Arts et Métiers, Paris, where it rests)

of the instrument were used for the measurement of electrical attraction and of magnetic repulsion and attraction. In 1784 Coulomb read a paper on his invention to the Academy of Sciences. This and subsequent papers were published in the *Mémoires* of that learned body. Already a member of the academy, Coulomb was now decorated as a knight of St. Louis, and in 1795 he was elected a member of the newly formed Institute of France. He was one of the leading physicists of the eighteenth century. His contribution to the knowledge of magnetism was greater than that of anyone else during the century. In electricity, his name has been immortalized as a term of definition for the practical unit of quantity.[30]

Early in the Revolution a micrometer was invented by a mechanical engineer named Richer, and after its approval by the Bureau of Consultation of the Arts and Trades he was awarded 6,000 livres by the Legislative Assembly (1792). The instrument, described as "very ingenous and very original" and as having "a very high degree of utility in mathematics and physics," could divide a ligne (a twelfth of an inch) into 1,200 equal parts. It was thus an instrument of high precision.[31]

To the English goes the credit for constructing the first iron bridges, in the late 1700's, but in 1755 an engineer named Garrin set out to construct one over the Rhône or the Saône at Lyons. It was to consist of three arches of twenty-five meters each, and only after one of them was mounted was the project altered because of the expense and the bridge completed in wood.[32]

---

[30] Wolf, *History of Science*, 245-49, 269-71; *Nouvelle biographie générale*, XII, 167-68; *Les membres et les correspondants de l'Académie Royale des Sciences, 1660-1793* (Paris, 1931), 59.

Coulomb, born into a family of magistrates in Angoulême, studied in Paris and then served in the army in Martinique for three years. Returning to Paris, he made valuable contacts with savants and men of influence. In July, 1774, he was named by the Academy of Sciences assistant to the Abbé Bossut, a mathematician and physicist who was interested in hydrodynamics. In the late 1770's Coulomb was sent by the government to Brittany to investigate the feasibility of digging a system of canals. When he returned opposing the project because of the huge outlay of money necessary, he was imprisoned in the Abbaye on the order of the minister and there held apparently for some time; yet he held firm in his opposition to the canals. He was one of ablest and most prolific contributors to the memoirs of the academy in the latter half of the century.

[31] Tuetey, *Répertoire générale*, VII, 266. The ligne was a twelfth of a pouce (1.066 inches) under the Old Regime.

[32] Arthur Vierendeel, *Esquisse d'une histoire de la technique* (2 vols., Brus-

To the English also must go the credit for the original invention of means for ventilating ships and buildings, but the French added improvements. The first of these instruments, invented by the English scientist Stephen Hales, was used on the *Sorbay* in the 1740's. It consisted of a box four feet long, twenty inches wide, and twelve inches high, encasing a bellows. Two men were required to operate it, and it had a capacity for supplying more than 25,000 cubic feet of air an hour. Used for refreshing the air in the stuffy hold of the *Sorbay*, it renewed the entire volume of air in the hold fifteen times each hour. It was also used to ventilate prisons and granaries. The machine was described by Hales in a pamphlet in 1744, and shortly afterward an account appeared in the French *Journal de Trévoux* of April, 1751. It was depicted as an instrument that could be of value to vessels transporting Negro slaves from the Guinea Coast.

As might be expected, this invention led to imitations and improvements. Two machines described as improvements were made by the Frenchman Pommier, a highway engineer, and approved by the Academy of Sciences in 1748 and 1752 respectively. The former, placed at the Hôtel des Invalides, was larger than Hales' ventilator, being four feet square and two feet high. Where Hales' instrument had vents for incoming and outgoing air on the same side, that at the Invalides had vents for incoming air on one side and vents for outgoing air on the other. A committee of four from the Academy of Sciences examined the operation of the instrument. A hall at the Invalides was filled with smoke almost to the point of suffocation. Then the ventilator was put to work, and after twelve minutes and 550 strokes of the ventilator handle, there remained insufficient smoke to inconvenience breathing. The ventilator of 1752 was yet more effective.[33]

In 1784 ventilators were in successful use in French military hospitals, in naval and merchant ships, at the Hôtel-Dieu of Paris, and at the depot of mendicity at Saint-Denis. Their expense was borne by the government. The machine at the Hôtel-Dieu

---

sels and Paris, 1921), I, 363, 366; Wolf, *History of Science*, 560-62.

[33] *Machines approuvées par l'Académie*, VII, 379-84, 413-14; *D.N.B.*,

VIII, 919. Martin Triewald, a Swedish captain, about the same time invented the device independent of Hales.

was boxlike in shape, five feet high, four feet wide, and three feet high, with two bellows and several vents.[34]

Other methods of ventilation were soon offered. Duhamel de Monceau, one of four appointed in 1748 by the Academy of Sciences to examine Pommier's first ventilator, is credited with suggesting in 1759 that the hold of ships be ventilated by means of kitchen stoves and flares. In 1767 Genneti proposed the ventilation in a similar manner of homes and hospitals by a channel of air admitted in the building on the ground floor, passed to the upper floors, and carried from the building by the chimney. This method, already suggested for ships by the Englishman Samuel Sutton in 1749, was a cheaper, marked improvement over that requiring bellows.[35]

A system of heating homes by the circulation of hot water through pipes was invented in 1777 by the architect Bonnemain, who introduced it that year in the Château du Pecq, near Saint-Germain-en-Laye. It worked successfully and remained in service for several decades. The temperature could be regulated by turning a small iron tube attached to a lead bar in the hot-water pipe. The water was heated by a stove on the ground floor and was carried through a circuit of pipes from room to room and from floor to floor.[36]

A device to prevent chimneys from smoking was invented early in the century and approved by the Academy of Sciences in 1715. It consisted of a rectangular piece of tin or sheet iron held in the middle by a pivot rod at the top of the chimney. From whatever quarter the wind blew, the sheet iron *(tôle)* extending above the chimney would be pushed by the force of the wind in the opposite direction, and this would aid the draft.[37]

Late in the century Lhomond, the Parisian physicist, invented a mechanism for controlling the heat of fires in coal grates. It consisted of metal sheets which worked in a frame like window sashes, each with its counterweight so that the shutter could be

---

[34] *Documents des hôpitaux de Paris*, II, 154-55. For a new ventilator by Teillard in 1790, see Tuetey, *Répertoire générale*, III, 538.

[35] Figuier, *Merveilles*, IV, 360-61; *Histoire de l'Académie, 1780*, p. 114. Figuier illustrates Genneti's method.

[36] A description and illustration are given in Figuier, *Merveilles*, IV, 315-16.

[37] *Machines approuvées par l'Académie*, III, 47-48. The inventor did not claim that his device would bring absolute, but only relative, freedom from smoking.

raised or lowered easily. When lowered, the draft in the fire was increased enormously. The metallic sheets could be lowered to the point of excluding view of the grate completely, an advantage for appearance of the room when fires were not needed. The invention proved popular and by the mid-nineteenth century was in common usage.[38]

In the early part of the century an improvement in the bellows was made by one Teral and approved by the Academy of Sciences in 1729. Where the bellows in common use was fashioned and operated on the principle of the lever, the bellows by Teral was controlled by a hand-turned wheel. To the end of this fan was attached a chain which ran to a large wheel on a nearby tripod. As the wheel was turned, the chain turned the fan and forced air through the bellows.[39] This form of bellows, with improvements, has continued in use into the twentieth century.

Eighteenth-century France saw much experimentation, and some success, with incubation of eggs. The priests of Isis in ancient Egypt practised incubation in ovens, but they kept secret the details of the art. Around 1588 the Italian Jean Baptiste Porta constructed at Naples an incubator designed after what he knew of ancient Egyptian practice, and attained some success. Rumor of what he was doing led to his acquiring the reputation of a sorcerer and attracting the attention of the Inquisition. As a safety measure Porta gave up his experiments. Time lapsed, and in the early eighteenth century Réaumur, reportedly without knowing of Porta's work, gave attention to this matter. His thermometer, one of the best of that day, was devised to check the temperature of his oven. After much experimentation, Réaumur attained some success at incubation.[40]

Much attention was given to the subject in mid-eighteenth-century France. Various persons, including the well-known physicist Abbé Nollet, made experiments. Some dreamed of riches to be made. Réaumur thought of the reduction in the price of fowls to Parisians, who had to depend so largely upon distant rural regions for their supply, and he hoped to inaugurate the day dreamed of by Henry IV when every peasant would have a

---

[38] Figuier, *Merveilles*, III, 263.
[39] *Machines approuvées par l'Académie*, V, 93-94.
[40] Paul Devaux, "L'incubation artificielle," in *Nouvelle revue*, LXXVII (1892), 570, 572-74.

chicken in his pot on Sunday. King Louis XV himself became interested in the enterprise and took pleasure in assisting the chicks to extricate themselves from their shells. And not only chicks, but also ducks, pheasants, quail, guinea fowl, and pea fowl were hatched in the ovens. Réaumur even experimented with ostrich eggs which he obtained from the zoo in Versailles.[41]

Greater success, however, was obtained by the Abbé Copineau on the eve of the Revolution, after he had experimented at much personal expense for two decades, and after he had read of Porta's work. Until the very last he met discouraging results, obtaining success only when he placed the eggs in a basket suspended by a wire within the smaller of two concentric cylinders, between which was a column of water, and under them an alcohol lamp.[42] Numerous eggs were hatched by this means, yet the instrument was not turned to commercial profit nor apparently was it approved by the Academy of Sciences. During the period of the Revolution experimentation with the incubator was carried further by Bonnemain and François Joseph Bralle (1750-1832), the former a physicist of Nanterre, the latter a prominent highway engineer.[43] In the *Moniteur* for January 19, 1792, there appeared a lengthy announcement of experimentation on an incubator and the public invitation to contribute 400,000 livres in shares of 500 livres for commercial exploitation of the invention. It was estimated that at least 400,000 fowls could be produced each year.[44]

At least three Frenchmen wrought improvements on the adding machine of Blaise Pascal, one of the notable inventions of the seventeenth century. One was Lépine, whose machine was approved by the Academy of Sciences in 1725. Its remarkable feature was a *sautoir* carrying figures from tens, hundreds, and thousands to the decimals next higher. Hillerin de Boistissandeau followed him with three similar machines, none of which had exceptional merit. In 1751, however, the Academy of Sciences approved a machine by the Bordelais Jew, Jacob Pereire (1715-

[41] Torlais, *Esprit encyclopédique*, 304-10.
[42] Devaux, "Incubation artificielle," 574-76.
[43] *Ibid.*, 576-77; *Nouvelle biographie générale*, VII, 228-29; *Grande encyclopédie*, VII, 979.

[44] *Moniteur universel*, XI, 152. Those wishing to subscribe were directed to address themselves to the home of Dufouleur, a notary. Whether or not Dufouleur was the inventor is not indicated.

1780), a famous worker with deaf-mutes, who invented his machine for use in teaching his deaf-mute pupils. It is described as "simple" but "ingenious." Calculating machines also absorbed the attention of certain gifted Germans and Englishmen. Among the Germans, Jacob Lerpold (1727), Matthew Hahn (about 1779), and J. H. Müller (1784) devised intricate machines of this type, while the English Viscount Mahon, afterward Earl of Stanhope, produced two of merit (1775, 1777).[45]

The safety lock, an old device, was improved by Sandos Le Gendre, an aged watch mechanic whom the Legislative Assembly in 1792 rewarded with a gift of 500 livres after a favorable report from the Assembly's Bureau of Consultation of the Arts and Trades and the Academy of Sciences. The only really effective lock of the century, however, was made by Joseph Bramah of England (1784). Bramah offered £200 to anyone who could pick it, and not until 1851 was it done, by Alfred Hobbs, an American, who labored fifty hours in doing so.[46]

In probably no other field of technology did careful French workmanship and ingenuity reveal itself to so great an extent as in horology during the 1700's. It was a century in which great developments in the making of watches, clocks, and chronometers were taking place, notably by the English, the Swiss, and the French. In this development a score or two expert French watchmakers took part, and the Academy of Sciences approved some three or four dozen of their improvements in horology. Many of these "inventions" represented novel forms and uses of the pendulum; others, new forms of escapement; many, different forms of dial, with hours, minutes, seconds, days, and months of the year indicated; various repeating watches; a number with planispheres attached; and still other features. The improvements were so minute and intricate in character that only trained watchmakers could describe them, and only watchmakers understand the descriptions. Certain of the foremost inventors in this field, however, should be sketched briefly.

[45] Wolf, *History of Science*, 655-60; Ernest La Rochelle, *Jacob Rodrigues Pereire, premier instituteur des sourds-muets en France: sa vie et ses travaux* (Paris, 1882), 77-81. Both books describe Pereire's machine in great detail. Addition and subtraction even in fractions was possible with this machine.

[46] Tuetey, *Répertoire générale*, VII, 269; Doolittle, *Inventions*, 424-25; Kaempffert, *American Invention*, II, 315.

In this group the family of Leroy stood forth pre-eminently, furnishing two of the most noted horological inventors of the century—Julian (1686-1759) and Pierre (1717-1785). Julian, born at Tours, moved to Paris, where he acquired celebrity as a watchmaker through decreasing the size of watches, making them more compact, removing sources of friction for the parts, and increasing precision.[47] His skill was praised by Voltaire. His eldest son Pierre entered a contest in 1763 for a prize offered by the Academy of Sciences for the chronometer which showed the greatest accuracy at sea. He met competition from Berthoud, another gifted horologer, and emerged the winner. His chronometer, however, did not equal the accuracy of one by John Harrison (1693-1776), which won the prize of £20,000 offered by the British government about this time. On a voyage to the West Indies Harrison's chronometer lost only five seconds.[48]

Leroy's chronometers (for he presented two) did not perform so well. They were carried on a voyage of forty-six days in the Channel and North Sea; at the end, one showed a loss of seven minutes, the other of thirty-eight. The next year on another voyage they showed up better.[49] Leroy's brilliant conceptions paved the way for the future development of the chronometer. Leroy made observations on the specific lengths needed for balance springs, he used a bimetallic balance of steel and brass to compensate for changes in temperature, and he devised a new and more efficient type of escapement. The French government recognized his skill with a pension of 1,200 livres, and on his death in 1785 continued half that sum to his widow.[50]

Other outstanding inventors among French horologers during the century were Henry Sully, a skilled Englishman brought to France by John Law, the inventor of an escapement and a chrono-

[47] *Nouvelle biographie générale*, XXX, 889. An escapement invented by Julian Leroy is described in *Machines approuvées par l'Académie*, VI, 83-84. A popular account of developments in horology in France during the century may be found in Franklin, *Vie privée*, IV, 123-61.

[48] See its description and illustration in Wolf, *History of Science*, 156. This was the fourth, the most compact, and the most accurate chronometer invented by Harrison.

[49] Usher, *History of Inventions*, 286-88, 290-91; *Nouvelle biographie générale*, XXX, 890.

[50] *Archives parlementaires*, XIV, 208; Wolf, *History of Science*, 157. His brother Jean Baptiste Leroy was a distinguished physicist and a member of the Academy of Sciences. He invented a machine for creating positive and

meter; Antoine Thiout (1692-1767), the author of a treatise on watchmaking and the inventor of a lathe and a new form of pendulum; Jean Romilly (1714-1796), a Genevan of French Protestant extraction, who settled at Paris and invented a watch capable of running a year without rewinding and wrote the articles on horology for the *Encyclopédie* of Diderot and D'Alembert; and Ferdinand Berthoud (1729-1807), another Swiss who came to Paris and there spent most of his life making chronometers with much variety in design and brilliance in conception. Next to Pierre Leroy, he was probably the cleverest French horologer of the eighteenth century.[51]

Also among the horologers was Beaumarchais (1732-1799), who followed his father into watchmaking and as a young man invented a new type of escapement which he naively showed to an older and distinguished watchmaker in Paris named Lepaute. Lepaute proceeded to appropriate the idea as his own and applied to the Academy of Sciences for its approval. Beaumarchais, however, did not stand by idly; in some published letters he attacked Lepaute severely and called upon the academy for a careful examination of his claims. A committee of two was appointed, who finally supported Beaumarchais, to whom was given the honor. The affair ran from 1752 to 1754.[52]

Altogether French clockmakers and watchmakers during the century made a notable contribution toward improvement in their field.[53] Governmental interest in the matter was shown both in Britain and France in an endeavor to arrive at better methods of determining longitude at sea.[54]

French scientists and opticians contributed a number of improvements to telescopy and microscopy, although in these fields English instrument-makers held pre-eminence in the eighteenth century. It was a century when remarkable strides were being

---

negative electrical charges and improved the lightning rod and the areometer. In 1772 he, too, was pensioned by the government.

[51] Usher, *History of Inventions*, 288, 327; Wolf, *History of Science*, 158; *Nouvelle biographie générale*, XLI, 587; ibid., XLV, 201; *Mémoires de l'Institut*, II, 122.

[52] The matter is described in great detail in Louis de Loménie, *Beaumarchais et son temps. Etudes sur la société en France au XVIIIe siècle, d'après des documents inédits* (2d ed., 2 vols., Paris, 1858), I, 76-80.

[53] Usher, *History of Inventions*, 327; Delambre, *Sciences mathématiques*, 215, 235.

[54] Bertrand, *Académie des Sciences*, 192-97.

made in the power, the complexity, and the adaptability of the telescope and the microscope. Among the French improvements of these instruments was the invention of a portable transit telescope by Jacques Eugène d'Allonville, the Chevalier de Louville (1671-1732), for the study of stars and planets as they rose above the horizon. It turned about a pivot or axis which itself was a horizontal telescope and served as a sort of spirit level to enable the observer to tell (through a mirror) if the instrument retained correct adjustment after being set up with a plumb line. The Academy of Sciences approved it in 1719. Louville was an aristocrat born near Chartres who fought first in the French navy and afterward in the army from about 1690 until 1713, when the War of the Spanish Succession ended. Then he devoted himself to the study of astronomy; in 1714 he was made a member of the Academy of Sciences. In 1717 he moved to a country home near Orléans, where he carried on his observations and research.[55]

Another contribution to telescopy was made by Pierre Bouguer (1698-1758) in his invention of the heliometer (1748). With this instrument one could measure small angles with great exactitude. Because it was first used to calculate the diameter of the sun, it came to be designated *heliometer,* but it was used afterward for many problems. In the nineteenth century, for example, the German astronomer Bessel determined for the first time the distance between the earth and a fixed star by means of a heliometer. The device is a screw micrometer with two lenses or lens segments attached to the object end of a telescope. An image is conveyed by each lens to the observer at the eyepiece. By turning the screw of the micrometer the lenses (and the images) can be brought together, or vice versa, as desired. The observer then records from the scale and dial of the micrometer the number of turns of the screw. In short, this device serves the observer as though he were using two telescopes, with means of calculating slight angular differences between them. This was an exceedingly useful invention, and later, in 1754, it was improved to a slight degree by Joseph Jérôme de Lalande (1732-1807), another celebrated French astronomer. Bouguer, the inventor of this instrument, was a native of Brittany, the son of a professor

[55] Wolf, *History of Science,* 132; *Nouvelle biographie générale,* XVI, 56-57.

of hydrography (by whom he was trained in the sciences), and after 1731 an associate member of the Academy of Sciences. In 1736 he accompanied La Condamine on his celebrated expedition to Peru for study of the shape of the earth.[56]

New types of reflecting telescopes invented by Le Maire and Claude Simeon Passement (1702-1769) were approved by the Academy of Sciences in 1732 and 1746 respectively. The latter also made a number of improvements in the microscope.[57]

Improved lenses were made by several men. Edme Sebastien Jeaurat (1724-1803), a prominent Parisian astronomer, produced a diplantidian telescopic lens that presented two images of the same object to the observer, one true and the other reversed. It was approved by the Academy of Sciences in 1779.[58] Carrochez in 1792 was awarded 6,000 livres by the Legislative Assembly for perfecting achromatic telescopic lenses and for improving telescopes by equipping them with platinum mirrors.[59] Already at this time a seven-foot telescopic lens had been made in France. The great problem for opticians in France in the last three decades of the century was to make flint glass, which only the English could make and which the French could obtain from them only in small quantity at high cost. The secret to the production of this glass, which had a lead content and was much better for telescopic and microscopic lenses than crown glass, was discovered at the outset of the nineteenth century by Louis Huette (1756-1805), a celebrated optician at Nantes, and also by Guinand of Switzerland. Born at Rennes, the son of a woodworker, Huette had left home at fifteen and for approximately eighteen years had traveled widely over Europe and the Near East and worked in various places before returning to France, where he made a number of inventions before his death. Among these was the improvement of the microscope in 1794 by the use of some achromatic lenses which he had made.[60]

---

[56] Wolf, *History of Science*, 143-44; *Nouvelle biographie générale*, III, 909-10. Wolf has an illustration of Bouguer's heliometer.

[57] *Machines approuvées par l'Académie*, VI, 61-65; *ibid.*, VII, 341-60; *Nouvelle biographie générale*, XX, 304-305.

[58] *Histoire de l'Académie*, 1779, pp. 36-37; *Nouvelle biographie générale*, XXV, 607-608.

[59] Tuetey, *Répertoire générale*, VI, 263.

[60] *Nouvelle biographie générale*, XIII, 395-96; *Encyclopédie méthodique*, IV, 252, 262.

Among the instruments of precision receiving improvements was the surveyor's level, which both Huette and the Abbé Soumille bettered. That by the Abbé Soumille, approved by the Academy of Sciences in 1737, was based on a pendulum and was sensitive to the slightest horizontal deviations. The instrument by Huette (completed in 1802), used both in astronomy and surveying, made use of the air bubble for determination of the horizontal position.[61]

The improvement of tools for engraving stones was made in the 1750's by Pierre Joseph de Rivaz (1711-1772) and in the 1790's by Philippe de Girard.[62] Claim to the invention of a better method of engraving paper money (assignats) and of coining gold and silver was made in 1790 by a Parisian metal engraver named Chipart. He offered to sell his rights to the government for 200,000 livres, and this high sum the National Assembly agreed to pay on condition that the claims of the inventor passed muster before an examining committee of three members of the Assembly and four members of the Academy of Sciences. Whether the committee's report was favorable is not indicated. Interest in the invention at such a price, however, would suggest that the French government was having trouble with counterfeiters.[63]

In the field of printing, a portable press fashioned of steel and copper was invented in the late 1770's by the Abbé Rochon, a friend of Turgot and the Abbé Morellet. This machine was operated by a single workman and was superior to any existing press. The quality of its work was so high as to approach engraving.[64]

More significant was the improvement of the stereotyping process late in the century by Firmin Didot (1764-1836), a member of the celebrated printing dynasty of Paris, several of whose members made contributions to the clarity and beauty of type during the century. The printing done by this family was a

[61] *Histoire de l'Académie*, 1737, p. 109; *Nouvelle biographie générale*, XIII, 395-96.
[62] *Nouvelle biographie générale*, XLII, 338; Ballot, "Girard," 145-46.
[63] *Archives parlementaires*, XIX, 495-96.
[64] Letter to the Earl of Shelburne, August 25, 1779, in André Morellet, *Lettres de l'Abbé Morellet de l'Académie française à Lord Shelburne, depuis Marquis de Lansdowne, 1772-1803*, ed. by Edmond Fitzmaurice (Paris, 1898), 168-70. See also Turgot, *Oeuvres*, V, 650; *Nouvelle biographie générale*, XLII, 466-68.

fine art, at least the equal of that of the Baskerville Press, and to a journalist in 1796 it seemed that no further improvement was possible. Firmin Didot has repeatedly been called the inventor of stereotype. Actually it was used in the period 1725-1750 by the Scot William Ged and his son James, and it appears that it was used even in late seventeenth-century Paris. Firmin Didot, however, made a notable improvement in the process and put it on a practical basis. He was made a knight of the Legion of Honor and later was elected to the Chamber of Deputies.[65]

The fountain pen was a French invention of the eighteenth century. The *Moniteur Universel* in two of its issues in 1790 (October 17 and December 3) carried accounts of this new instrument which was described as capable of writing several hours without need of refilling and of being a great aid to travelers and those taking notes at public meetings. It was sold cheaply—a penstaff, in ivory or ebony, with six pens and a bottle of ink for six livres. The inventor was Coulon de Thévenot (1754-1814), who called his invention a "pen without end."[66]

Coulon de Thévenot was also the inventor of a system of shorthand or tachygraphy which in 1783 won the approval of the Academy of Toulouse and in 1787 that of the Academy of Sciences in Paris. For ten years he had labored at his task, being aided in its later stages by certain suggestions from the committee appointed by the Academy of Sciences to investigate it. In 1787 he published a booklet setting forth the details of his system, and during succeeding years he taught it to scores of students from all parts of France and all walks of life. Some were newsmen; others, deputies of the National Assembly and subsequent legislative bodies. Coulon himself was made royal tachygrapher in 1787, and in 1789 secretary to the chief of staff of the National Guard of Paris. During the 1790's he held a number of important secretarial posts, and his system of shorthand made a remarkable showing. This capable man deserved a better fate. In the early

---

[65] *Grande encyclopédie*, XI, 670; *Nouvelle biographie générale*, XIII, 115; Paris *Mercure français*, 20 fructidor an IV, 81-85.

[66] *Moniteur universel*, VI, 140, 532; J. F. Coulon de Thévenot, *L'art d'écrire aussi vite qu'on parle; ou, la tachygraphie française, dégagée de toute équivoque* . . . (New ed., Paris, 179 ?), 21, 42.

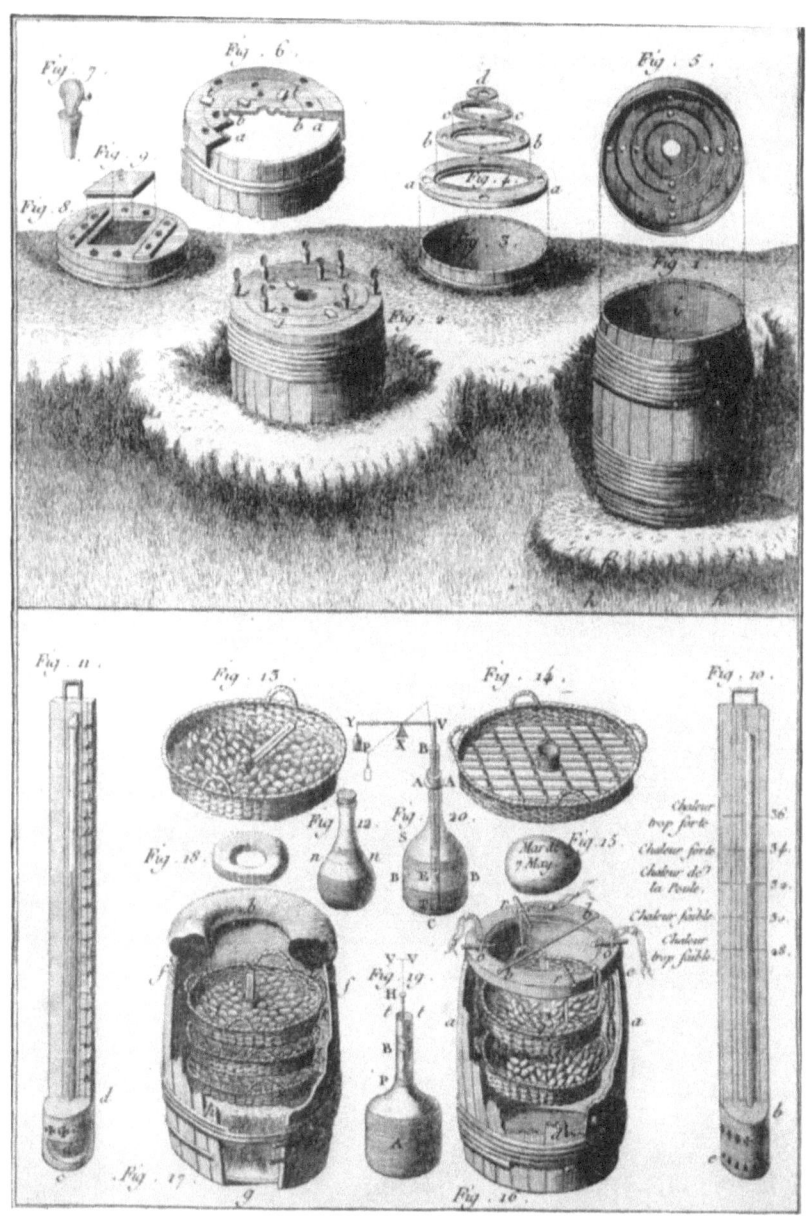

AN INCUBATOR FOR HATCHING CHICKENS

The diagrams depict exterior and interior views of a late eighteenth-century incubator.

(Illustration from *Encyclopédie méthodique, Arts et métiers*, XIX, pl. 33, courtesy of the University of Kentucky Library)

OTHER MECHANICAL DEVICES

1800's he enlisted in the army; after the disastrous battle of Leipzig (October, 1813) he was captured by the Cossacks, robbed of his clothing, and left to die of exposure.[67]

In music, there were several inventions. Among them was a new form of harpsichord invented by a certain Cuisinié, approved by the Academy of Sciences in 1708.[68] It was followed in 1716 by the invention of an improved model by Marius, in which a keyboard with hammers was substituted for the jack and quill. The Academy of Sciences, approving it in 1716, described it as giving "stronger and more beautiful tones" than previous harpsichords and as being operated solely by touch. The instrument approached, if it was not in fact, the pianoforte.[69] An organ without bellows and an improved flute were invented around 1791 by a former Benedictine monk named Luxeul. The flute had a tin mouthpiece and mechanical devices for opening and closing the stops.[70]

The most interesting eighteenth-century French musical invention was the ocular harpsichord by the Jesuit Father Louis Bertrand Castel (1688-1757), a native of Montpellier and a professor at the Collège de Louis-le-Grand. Developing a suggestion made by the Jesuit Kircher that a relation exists between sound and light waves, he labored from 1730 to 1754 to invent and perfect a mechanism that would illustrate this. Montesquieu, the Duke de Huescar, and others contributed heavily in a financial way. Some scoffed, but the public in general was sympathetic and greatly applauded Castel's stages of success. Among his admirers was Diderot, who inserted in the *Encyclopédie* an article on the "Clavecin oculaire" and mentioned it in other writings. Unfortunately no clear description has been left of the instrument. Above the harpsichord, however, was a box or cabinet with numerous

---

[67] Coulon, *Art d'écrire*, 11, 20-21; *Histoire de l'Académie*, 1787, pp. 9-18; *Nouvelle biographie générale*, XII, 171-72; *Grande encyclopédie*, XIII, 51.

The investigating committee expressed its regret that the inventor could not use his own system with facility. This arose partly from the fact that from first to last Coulon had compiled six drafts or variants of his system. The system approved and published was actually his first draft revised.

[68] A description and drawing are in *Machines approuvées par l'Académie*, II, 155-56.

[69] *Histoire de l'Académie*, 1716, p. 77.

[70] *Moniteur universel*, IX, 62. Luxeul's flute appears to have been a forerunner of the Boehm flute, invented in 1832.

lights and a window made of scores of pieces of colored glass. Wires or cords of silk connected the keys with various colors, so that when a key of the harpsichord was pressed a corresponding color instantly flashed to view in the glass compartment or window. It is not surprising that spectators were enamored with the instrument and that all Europe quickly heard of it, the precursor to the kaleidophon and the clavilux of our own day. The invention, however, was bizarre and ephemeral and did not have the significance that Castel hoped for it.[71]

The parasol, an earlier invention which had appeared in Italy probably as early as 1578 and had been seen in Paris in 1622, underwent improvement and in 1705 the Academy of Sciences approved two forms of parasol with new features, and in 1709 a third form, all by Marius (d. 1720). At least two of these embodied many characteristics of twentieth-century parasols and umbrellas.[72]

Among the great variety of French mechanical inventions of the century can also be mentioned the diver's helmet with respirator, designed to recover lost treasure from sunken vessels. The idea of this contrivance is said to have occurred first to a wigmaker or barber in 1772 on news of a gold-laden ship sinking off the Spanish coast. He communicated his idea to Périer, who turned this fanciful device into a reality and with it is said to have descended into the Seine at Paris and there worked before a large crowd of witnesses. Later he experimented with it in ocean waters and retrieved two anchors from a depth of fifty-two feet.[73]

An air-filled vest for supporting a man from sinking at sea was invented late in the century by a surgeon named Le Conte, who

[71] Donald S. Schier, *Louis Bertrand Castel, Anti-Newtonian Scientist* (Cedar Rapids, Iowa, 1941), 135-96. I am indebted to Professor Edward D. Seeber of Indiana University for drawing my attention to this invention.

[72] *Machines approuvées par l'Académie*, II, 87-91, 161-62; Edouard Fournier, *Le vieux-neuf: histoire ancienne des inventions et découvertes modernes* (2 vols., Paris, 1859), II, 211 n.

[73] Lacroix, *XVIIIme siècle*, 64-65.

Later, in December, 1790, the National Assembly gave attention to the alleged invention of a similar "hydraulic machine" by Augier. In the discussion a deputy named Boussion reported that twenty years previously an invention of this type had been rejected by the government because of its lack of value. He proposed a careful examination of Augier's device. The matter was not reported again. *Archives parlementaires*, XXI, 642.

for this and certain medical discoveries was awarded a grant of 3,000 livres by the Legislative Assembly in 1792.[74]

Last and probably the greatest of the French inventions of the century was the metric system created and adopted in the 1790's. As early as 1585 a decimal system in coinage, weights, and measures had been suggested by Simon Stevin of Antwerp in his notable little book entitled *The Decimal*. Shortly before the Revolution various thinkers including Turgot had renewed the suggestion.[75] Discontent with the weights and measures in vogue in France—weights and measures which lacked uniformity throughout the country—was expressed in many of the cahiers in early 1789.[76]

The National Assembly was accordingly carrying out a mandate from the people when in May, 1790, it passed an enactment to bring about a new system of weights and measures and asked the Academy of Sciences to report upon such a plan. It was recommended that the unit of length be one ten-millionth part of the meridional distance from the North Pole to the equator, through Paris. This would be roughly equivalent to the ell, the yard, and the braccio (linear units for France, Britain, and Italy). A list of names was suggested for the decimal multiples and divisions of this unit. Likewise there was to be a unit of weight— the mass of a cubic centimeter of water at 4° Centigrade, its temperature of maximum density. The whole system was to be named after the Greek word for measure, *metron*.[77]

The report was adopted in March, 1791. The meridional quadrant was taken as the basis of linear measurement, and the Academy of Sciences was directed to undertake the measurement of a section of this quadrant—from Dunkirk to Barcelona. The academy organized committees for this difficult work. The ten-

---

[74] Tuetey, *Répertoire générale*, VII, 267.

[75] Preserved Smith, *A History of Modern Culture* (2 vols., New York, 1930-1934), 93; A. R. J. Turgot, *The Life and Writings of Turgot, Comptroller-General of France, 1774-6*, ed. by W. Walker Stephens (London, 1895), 112; Caullery, *French Science*, 66-67.

[76] In the sénéchaussée of Cahors alone, two communities (Saint-Martin-Labouval and Valroufié) demanded uniformity of weights and measures for all France. *Cahiers de doléances de la sénéchaussée de Cahors pour les Etats Généraux de 1789*, ed. by Victor Fourastié (Cahors, 1908), 297, 351. Many other instances could be cited.

[77] Arthur E. Kennelly, *Vestiges of*

millionth part of the quadrant had to be shown accurately in meter bars, copies of which were to be distributed widely. A unit of weight also had to be carefully determined, and standards established for it.[78]

The Academy of Sciences was abolished in August, 1793, but in October the Institut de France was established in its stead and took over the work of determining values of the meter and the gram. Throughout the Reign of Terror and subsequent years the work of triangulation to determine the distance between Dunkirk and Barcelona continued.

In April, 1795, before measurement of this distance had been completed, the Convention decided to proceed with adoption of the metric system. Tentative units of measurement were established—the meter as the unit of length and the gram as the unit of weight. Prefixes for expressing multiple units, whether of weight or length, were borrowed from Greek numerals—*deca* (ten), *hecto* (hundred), *kilo* (thousand), and *myria* (ten thousand); fractions were to be designated in Latin terms—*deci* (tenth), *centi* (hundredth), and *milli* (thousandth).[79]

In June, 1799, four years after the legal adoption of the metric system by French government, the standard meter and gram were presented by a committee of the Institut de France to the Council of the Five Hundred. Since the gram was a small unit, the standard in weight was represented in the kilogram. Before deciding on the best metal to use, experiments were made with standard units in several metals; platinum was chosen as most resistant to variations in temperature. Standard units were deposited by Napoleon at the Observatory for safekeeping.[80]

In the last phases of this work certain foreign scientists on invitation co-operated with the French. Indeed, as early as 1790 the French government had requested the British to join in its plan to set up a new system of weights and measures, without result.[81] In 1798 it invited several European governments to send delegates to Paris to participate in the fixing of the proper values

---

Pre-Metric Weights and Measures Persisting in Metric-System Europe, 1926-1927 (New York, 1928), 13.
[78] *Ibid.*, 14-15.
[79] *Ibid.*, 15-16.
[80] *Ibid.*, 17-19.
[81] Bertrand, *Académie des Sciences*, 407-408.

for the standard meter and the standard kilogram. Several responded. The joint commission that came into being reported on the problem in 1799.[82]

Even after 1799 the metric system was slow in making headway in France. The law did not designate when it was to come into use, and did not provide machinery for its enforcement. Even in 1812 Napoleon made some concessions to the old system of weights and measures, although requiring the metric system to be taught exclusively in the government offices. This hybrid state of affairs proved unsatisfactory, however, and in 1837 the French parliament after considerable debate established the metric system as the sole standard. Evidently even this action was not enough, for in 1840 a penalty was established for its nonusage—ten francs for every violation. Thus the metric system did not completely win the field in France until after 1840. Adoption by other countries, except only for Switzerland which accepted some of its features as early as 1822, came in the latter part of the century.[83]

The metric system is decidedly the best system of weights and measures in the world today, and for it the world is indebted to French genius of the eighteenth century. It was probably the greatest French invention of that century, a tremendous labor-saving scheme; yet it brought no wealth to France.

The French in 1795 were guilty of one inconsistency in adoption of the decimal system. They did not apply it wholly in the revision of their coinage. They changed the old livre to the franc with virtually no difference in value, and they left the sol or sou at a twentieth of the franc. The denier (subdivision of the sou) happily was changed, with a slight decrease in value, to the fifth part of a sou and the hundredth part of the franc. Nevertheless, the French monetary system was improved.

---

[82] Kennelly, *Pre-Metric Weights*, 17, 19. Fourteen foreign and thirteen French scientists took part in fixing the standard units.

[83] *Ibid.*, 20. Spain adopted the metric system in 1860; Italy, in 1861; Germany and Portugal, in 1872; Austria, in 1876; Norway, in 1882; Servia, in 1883; Roumania, in 1884; Bulgaria, in 1892; Denmark, in 1912; Russia and Greece, in 1922.

CHAPTER X

Military Inventions

EIGHTEENTH-CENTURY FRANCE WAS A GREAT MILITARY power. She had one of the largest and most efficient armies of Europe; during the Revolutionary era her troops swept everything before them. Such an army was not created overnight. Invention, which was transforming all other phases of human development, also contributed greatly to the military growth of the nation.

Long before the days of Lazare Carnot, the genius of the Revolutionary army, France had produced some remarkable military tacticians and organizers in the Chevalier Folard (1669-1752), the Comte de Guibert (1743-1790), and Jean Baptiste Vaquette de Gribeauval (1715-1789). Folard had served in the War of the Spanish Succession (1702-1714) with such distinction as to acquire an international reputation. In 1724 he began publishing a number of writings, based on study of the ancients, in which he advocated a battle formation of troops similar in part to the old Greek phalanx. His army would consist of battalions of six hundred men, one hundred armed with eleven-foot pikes to constitute the vanguard and flanks of a loose square formation and the remainder, carrying guns, the inner core, arranged in a series of five files or columns with some space between to gain more flexibility of movement than the ancient phalanx had. Frederick the Great scoffed at this system but other strategists, such as Maurice de Saxe, saw in it points of excellence. The chief merit was its ability to stop cavalry attacks. It could also easily deploy into line when occasion demanded.[1]

Another provocative tactician of the century was the well-read Comte de Guibert, who in 1779 at the age of twenty-seven pub-

[1] Oliver Lyman Spaulding, Jr., Hoffman Nickerson, and John Womack Wright, *Warfare: A Study of Military Methods from the Earliest Times* (New York, 1925), 553; *Nouvelle biographie générale*, XVIII, 52-55.

lished his famous *Essai général de tactique,* a work so filled with radical ideas as to be regarded as not only military but even political. It was discussed throughout Europe and translated into various languages, including Persian. Guibert predicted that any European country which would revolutionize itself, setting up a firm government and a program of aggrandizement, would sweep all opposition before it. This of course happened shortly afterward in the French Revolution. Guibert followed the Marshal de Saxe in recommending that soldiers be allowed to fire at random rather than on the order for a volley. He went beyond Saxe in urging that all infantry units be light and mobile. He agreed with Pierre de Bourcet, another French tactician, in advocating rapid warfare, such as falling on the enemy's supply trains and cutting them off. Napoleon later adopted this feature. Guibert modestly attributed most of his ideas to earlier men and claimed to be primarily a compiler and publicist. A recent writer, however, has given him much credit for the victories of French armies during the Revolution.[2]

More important than either Folard or Guibert was Gribeauval, who for many years after 1776 was inspector general of French artillery. He introduced into the French army a large number of reforms, some of which had been suggested to him while he served during his younger days in the Prussian and Austrian armies. Interestingly enough, this foreign service was at the direction of the French government, which hoped to learn of improvements for its own military forces. Gribeauval had had twenty years' service in the French army when in 1752 he was chosen by the Comte d'Argenson, minister of war, to study the Prussian artillery. This mission he completed by 1756, when the Seven Years' War broke out. Then he entered the Austrian army, in which he served with distinction, in 1762 becoming a lieutenant field marshal. With peace he was summoned back to France by the Duc de Choiseul, minister of foreign affairs.

As inspector general of artillery, Gribeauval transformed the French artillery schools; he formed a corps of sappers; he improved the arms manufactories, forges, and foundries; he changed the artillery cannon to a smaller calibre, so gaining in mobility;

[2] Lynn Montross, *War through the Ages* (rev. ed., New York and London, 1946), 447-49.

he ordered the gun horses to be hitched in pairs instead of single file; he introduced a new gun carriage of his own invention for coastal defense; he altered the construction of cannon, abolishing the old powder chamber as obsolete and so reducing the quantity of gunpowder needed; and he adopted the tangent sight for greater accuracy in aiming. Possibly more significant than anything else he did, he standardized firearms and fieldpieces, so that they might be produced in large quantities with their parts interchangeable. Gribeauval's innovation proved of great value in the Revolutionary period when vast numbers of guns were needed. The French artillery is reported to have excelled all others in Europe, thanks to Gribeauval's improvements.[3]

During the Revolution the French made their most original contribution to modern warfare—the famous *levée en masse*, decreed by the National Convention August 23, 1793, early in the Reign of Terror when Republican armies in northern France had given way before the Austrians and Prussians and in various parts of the country royalists had risen in insurrection. As early as December, 1789, Dubois-Crancé, an army captain and later a general, had proposed to the National Assembly the adoption of military conscription, but the idea had aroused no interest.[4] After the outbreak of war in April, 1792, however, conscription gradually was instituted. The first step came on July 12, 1792, when the Convention ordered the drafting of 85,400 men, part to serve with the army and the rest with the national guards in defending the frontiers; then on February 26, 1793, a new levy of 300,000 men was ordered. This second order was shortly after the execution of Louis XVI, when it was ominous that the French would soon be faced in battle by other nations besides the Austrians and the Prussians, as indeed did happen in March. The act of February 26 stipulated that each department was to fill its quota with volunteers; only when these were insufficient were men between the ages of eighteen and forty to be drafted.[5]

[3] *Ibid.*, 445-46; Spaulding, *Warfare*, 568; *Nouvelle biographie générale*, XXII, 19-23.
[4] Hoffman Nickerson, *The Armed Horde, 1793-1939: A Study of the Rise, Survival and Decline of the Mass Army* (New York, 1940), 67-68.
[5] *Moniteur universel*, XIII, 138; *ibid.*, XV, 550.

These orders were but preliminaries to the *levée en masse*, about whose author nothing is known except that he was a representative of one of the provincial assemblies, who together with numerous other departmental officials had come to Paris for the August 10 celebration of the first anniversary of the overthrow of the monarchy. Many of these visitors lingered in the city after the festivities. On August 14 the unknown representative demanded in the Convention not only military but civilian conscription. The brief, stirring speech was vigorously applauded, and the Convention ordered its committee of public safety to present a report of prospective legislation on conscription. This report, made August 20, for two days was much debated. The president then returned it to the committee with the order for a new one embodying the demands made on the floor. The second report was read before the Convention August 23, 1793, by the committee spokesman Barère, and the proposed decree was immediately adopted.[6]

In stirring, staccato notes the act called for the drafting of all unmarried men between the ages of eighteen and twenty-five who were not engaged in agriculture. It ordered also the conscription of all other civilians, even old men, women, and children, for labor in the factories and in the fields, in the transport service and in the hospitals. Saddle horses were ordered seized as needed for cavalry service; draft horses, for the artillery. Firearms of all types were to be turned over to the government for the use of the troops. Public buildings were to be converted into barracks, public squares to be made the sites of factories. The dirt in basements and barns was to be treated with lye to obtain saltpeter. In short, all the nation's resources—human, animal, physical—were mobilized for war. This action, which appears to have resulted from the spirited, timely demand of the unnamed provincial representative, has been described as "one of the most memorable dates in the chronicles of war."[7]

---

[6] *Ibid.*, XVII, 410-11, 412, 444, 474-78; *Recueil des actes du comité de salut public avec la correspondance officielle des représentants en mission et le registre du conseil exécutif provisoire*, ed. by F. A. Aulard (26 vols., Paris, 1889-1923), VI, 72-76.

[7] Montross, *War through the Ages*, 452.

In military engineering the French also made certain inventions and innovations. A pontoon bridge for the rapid crossing of rivers was invented in 1710 by the gifted François Joseph Camus. Although floating bridges date at least from Caesar's use of one for crossing the Rhine in the first century B. C., Camus would not allow that the features of his bridge would be duplicated by mere coincidence in the plans for one submitted to the Academy of Sciences in 1773 by D'Herman. Yet D'Herman insisted that he had invented the bridge independently of Camus and had not known of the latter's exhibition at Bercy, near Paris, three years earlier. The Academy accepted his story and approved the bridge as the work of D'Herman, but mentioned the prior claim of Camus. This bridge was six feet wide, permitting three soldiers to march across it abreast, and it could be set up in 10 minutes, 35 seconds, as demonstrated on the large canal at Versailles in the presence of Louis XIV.[8]

A famous gun carriage for use in coastal artillery, credited to Gribeauval who was largely responsible for its adoption, was the invention of Claude François Berthelot (1718-1800). This poor youth rose by dint of hard work and his own ability to the professorship of mathematics in the Military School in Paris, where in 1763 he invented his gun carriage. That same year he installed one at the arsenal at Auxonne, and in 1764 a second at Strasbourg. In 1765 Gribeauval made a report for the ministry of war urging its adoption generally for coastal defense batteries, and Berthelot obtained a pension of 600 livres, paid until the Revolution.[9]

While Gribeauval was transforming French artillery, the Marquis de Montalembert was revising French ideas of engineering defense in a notable book, *Fortification perpendiculaire*, and his ideas prevailed in the late 1700's and early 1800's throughout Europe. He was the Vauban of the late eighteenth century. He advocated the construction of defense fortifications along polygonal lines, so that the besieged were able "to pour a superior fire into the works of a besieger."[10]

---

[8] *Machines approuvées par l'Académie*, III, 17-18.

[9] *Nouvelle biographie générale*, V, 703; *Biographie universelle*, LIII, 99-100.

[10] Montross, *War through the Ages*, 446; *Nouvelle biographie générale*, XXXVI, 85-87.

No account of French military innovations for this century can be written without mention of the bizarre floating batteries invented to reduce Gibraltar. The project, "a secret weapon" in the War of American Independence,[11] was conceived in 1780 by Jean Claude Eléonore Le Michaud, Chevalier d'Arçon (1733-1800), a military engineer trained at the college of Mezières. The French government adopted the scheme with enthusiasm and made plans for a thoroughgoing attack on Gibraltar in the late summer of 1782. Ten floating fortresses (eight of them commanded by Spaniards), a Spanish-French fleet, and an army of 40,000 men under the leadership of the Duc de Crillon were assembled for the undertaking.

All hopes were pinned on the floating fortresses which sailed within a half mile of the great rock and attacked it in the early morning of September 13, 1782. The fight lasted into the night and the next morning, but the shelling of the British by all land and sea forces of the Allies had little effect. Most of the supposedly fireproof floating fortresses were set afire by British hot shot during the night of the thirteenth and morning of the fourteenth, after their first fruitless efforts with cold shot, including a thirteen-inch ball that rebounded as if rubber from the sides of the floating batteries.

The floating fortresses were warships whose sides were reinforced by an outer casing of logs and an intermediate layer of sand and cork which was flooded with water before the engagement so as to withstand attack from combustible shells. Each battery, moreover, had a sloping roof (a bizarre feature for ships), formed by rope netting over which was a heavy covering of hides, to protect the crew. Through the roofing rose the masts carrying rigging and sails. Each battery was well equipped with cannon which fired through apertures in the sides. The batteries put up an excellent fight, and it is amazing that their significance

[11] Earl J. Hamilton, "War and Inflation in Spain, 1780-1800," in *Quarterly Journal of Economics*, LIX (1944-1945), 40 n. 6. See more extended accounts in Samuel Ancell, *A Journal of the Late and Important Blockade and Siege of Gibraltar, from the Twelfth of September 1779, to the Third Day of February 1783* . . . (3d ed., Edinburgh, 1786), 224-28; Frederic George Stephens, *A History of Gibraltar and Its Sieges* (2d ed., London, 1873), 272-80; John Drinkwater Bethune, *A History of the Siege of Gibraltar, 1779-1783* (new ed., London, 1905), 295-307.

for naval warfare was not appreciated in the years that followed.[12]

Few inventions in personal weapons were made, and these were largely developments of firearms already in use. Thus in 1704 the Academy of Sciences gave its approval to a breech-loading musket and in 1715 to a breech-loading cannon invented by De la Chaumette.[13] But breech-loading guns and cannons antedated the eighteenth century.[14] De la Chaumette's socket bayonet, approved by the academy in 1707, also had been anticipated in 1691 by the British general Mackay.[15]

A breech-loading rifled carbine was invented by Maurice de Saxe, the famous French marshal of the mid-century, but again this could only have been a development of earlier models.[16] Saxe also invented a one-pound light cannon, which he called an *amusette*.

In 1716 the Academy of Sciences approved two large water-propelled machines for drilling small arms barrels, built on plans submitted by Villons. Possibly they represented the independent work of Villons, yet it is of interest that in 1714 Jean Maritz, a Swiss founder, invented in his own country a method of boring cannon by means of a horizontal drill. Before this time cannon had been cast about an iron core covered with a layer of clay. Maritz cast the cannon solid and then bored it. The cannon was turned about a horizontal drill, rather than vice versa. At the invitation of the government Maritz headed the foundry at Lyons, and his method exclusively was soon used in France. He died in 1743 at Geneva, but his son became a naturalized Frenchman and carried on his father's ideas, receiving high honors and a considerable pension from the government, especially when the Spanish and Russian governments made bids for his services.[17]

---

[12] Details concerning the batteries are in Stephens, *History of Gibraltar*, 280.

[13] *Machines approuvées par l'Académie*, II, 79-80; *ibid.*, III, 53. In both instances a lever below the cartridge chamber could be pulled down and ball and powder inserted; then the lever was pushed back into place.

[14] Fournier, *Vieux-neuf*, I, 310-11 n. 2.

[15] *Machines approuvées par l'Académie*, II, 149-50; Figuier, *Merveilles*, III, 472. The bayonet had originated in the Basque country in 1641. In 1703, on the recommendation of Vauban, it was adopted for all French infantry units; but it had to be removed for firing. The socket bayonet fitted over the gun barrel and permitted firing at any time.

[16] Spaulding, *Warfare*, 554.

[17] Ballot, *Machinisme*, 421-22; *Machines approuvées par l'Académie*, III, 73-74; *Nouvelle biographie générale*, XXXIII, 807.

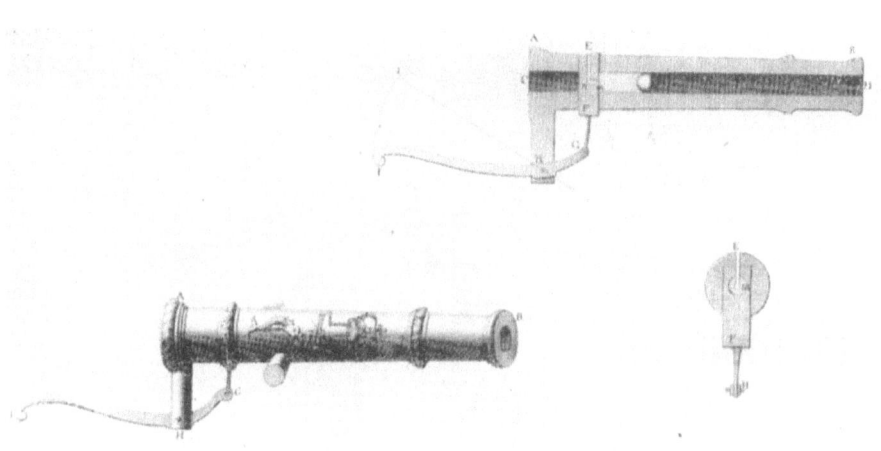

THE BREECH-LOADING CANNON OF DE LA CHAUMETTE

This weapon was patented in 1715.

(Illustration from *Machines et inventions approuvées par l'Académie royale des Sciences*, II, 156, courtesy of the Ohio State University Library)

The work of Maritz and Villons presaged the more celebrated and accurate cannon-boring drill of 1800.

Bernard Forest de Belidor (*ca.* 1698-1761) came forward in the mid-century with an invention he called "globes of compression." Later, during his service with the Austrian army in the Seven Years' War, Gribeauval worked at its improvement. These "globes of compression" appear to have been shrapnel, invention of which is attributed to the English Major Henry Shrapnel in 1784; in this case the French were first.[18]

Throughout the second half of the century numerous reports arose concerning the invention of "Greek fire" by one Frenchman or another.[19] The reports are gossipy and none too reliable. All have this in common—the French government was horrified at the idea of using so powerful, so vile, so unfair a weapon against an enemy and placed the inventor under the most strict orders never under any circumstances to reveal the secret to any man or any other country. When finally rumor came that it had been communicated by one of these French inventors to the British government, the suggestion arose in France for an international agreement to outlaw its use.

The stories began with the invention of an incendiary bullet in 1759. Madame de Genlis, apparently in error, stated that Torré experimented successfully on the canal at Versailles, and added that Louis XV bade him never to publish the secret. The king declared that France never could condescend to such a crime as using it against her enemies would be. Yet he gave Torré a *brevet d'invention* and a pension of 1,000 écus (3,000 livres).

[18] *Nouvelle biographie générale*, V, 197-98; *ibid.*, XXII, 21; Montross, *War through the Ages*, 515. Belidor was born at Catalonia. Orphaned when five months old, he was raised by an artillery officer. He worked under Lahire and Cassini at cartography, and eventually he became a professor in the artillery school at La Fère.

[19] The Byzantine Greeks of the Middle Ages made use of combustible materials in naval warfare; hence the name "Greek fire." These combustibles varied in content and in form. Their chief constituent appears to have been naphtha. Other ingredients probably were sulphur and quicklime. They were hurled against enemy ships sometimes as hand grenades thrown by sailors, sometimes as pots of liquid thrown by mechanical devices, and sometimes made by a gunner to slide down a long tube called a siphonator. On contact they either exploded or burst into flame. See an interesting paragraph on the subject by J. B. Bury in his edition of Edward Gibbon, *The History of the Decline and Fall of the Roman Empire* (2d ed., 7 vols., London, 1929), VI, 562.

Another story, by the writer Coste d'Arnobat, identified the inventor as Coste from Dauphinais, who died in 1772; and his pension, instead of 1,000 écus, was 2,000. The weapon was a wooden gun which fired 700 arrows covered with a composition which burst into flame when an object was hit.[20]

In 1786 an artillery officer named Bellegarde invented an incendiary bullet and told its secret to the British, after first giving it to the French army and navy departments.[21] In 1793, an artillery officer named Fabre, employed at La Fère, claimed that he was the actual inventor of the Bellegarde bullet. This time the French government took more interest in the matter, possibly because it was at war and fortune had not sufficiently smiled on French arms, and ordered 3,700 of the bullets for coastal protection. This action impelled a citizen named Prévost to urge the government to make certain whether the British had the bullets and if so to come to an agreement with them not to use the weapon.[22]

Some months later news of another inventor of Greek fire came from a strange quarter. At the meeting of the National Convention on 10 fructidor an II (August 27, 1794) a letter was read from the Rev. J. P. Coste, pastor of the French Protestant Church of Charleston, South Carolina, offering to the French government in its national emergency a shell of his invention that burst into violent flames at a prescribed distance, and that no vessel could resist. It could also be used to terrible effect against cavalry and infantry, and could be hurled against a wall, which would remain enflamed for a half hour. Its flame and its odor were enough to throw panic into any enemy. But more than that, Coste claimed that six French vessels of the line equipped with this new, devastating weapon could "attack all the navy of Europe in a single day and there would not remain a yawl in any port." He was by no means modest in his claims. He swore, moreover, to communicate his invention to no other nation, not even under pain of death. Here was a patriotic Frenchman and

---

[20] Fournier, *Vieux-neuf*, I, 256-58. Fournier identifies the man definitely as Dupré.

[21] Gabriel Vautier, "Les bullets incendiaires en 1793," in *Revue historique de la Révolution française*, IX (1918), 509.

[22] *Ibid.*, 510.

an ardent Revolutionary, though a citizen of the United States. His proposal, however, shocked even the Convention deputy Baraillon, who commented that the invention sounded like Greek fire to him—a device which even a tyrant like Louis XV refused to use when reported to him by a brilliant chemist named Delille. He proposed that the matter be sent for examination by the committees of war and public safety, which might decide whether it "could be employed usefully, without danger to humanity." To the committees it was sent and there apparently died.[23]

But not for long. In 1798 General L'Espinasse invented a large, circular bullet, consisting mainly of cotton treated with chemicals. Whatever it was fired against it set afire. Experiments were made with success at Lorient, but it was not used against the enemy, either out of humane consideration or French indifference.[24]

Another invention of the latter part of the century was a rapid-fire weapon for slaughter of the enemy. A rifle or musket that fired twenty-four times in succession without reloading, by a circular movement of its magazine, was approved by the Academy of Sciences in 1767. It was the invention of a father and son Bouillet at Saint-Etienne, in Forez. Tests showed that after eighteen rounds were fired in a minute and a half, the barrel became almost too hot to hold, and it was found advisable to wait two minutes before firing the remaining shots. The committee thought the gun might be fired without interruption by using a glove. The weapon had the advantage of weighing only seven pounds, while the soldier's musket of that day weighed eight.[25]

Other claimants to rapid-fire weapons came forward in 1792. One of them, Milot by name, was introduced before the National Assembly on April 14 as the inventor of "a machine of war" capable of firing fifty shots a minute. The Assembly paid him its respects and sent his device for examination to its committees

---

[23] *Moniteur universel*, XXI, 606. Baraillon was deputy from La Creuse and a physicist.

[24] Fournier, *Vieux-neuf*, I, 259. In a footnote Fournier says that Brun de Condamine in 1773 proposed the use of an incendiary bullet but that Louis XV would not hear to the idea; when Condamine decided to carry the secret elsewhere, he was placed in the Bastille until peace was signed in 1782.

[25] *Histoire de l'Académie*, 1767, p. 186.

on military affairs and education, where it seems to have been buried.[26] In August a mechanic named Renard was introduced before the same Assembly with the story that he was the author of "a thunderbolt of war" capable of firing ninety times in a second, operated by one man. This invention was sent for investigation to the commission of arms and the military committee, with like results.[27] A German named Gass, chief chemist at the porcelain factory at Sèvres, claimed the invention of a pistol which would fire six to fifteen bullets a second.[28]

A mechanic named Garnier asserted he had invented four machines which would make large and small caliber cartridges that required no biting or ramming, so facilitating greater speed in firing. Like the other reputed inventions, it was sent by the Assembly to a group of experts for examination and report.[29]

In October, 1793, the deputy Moreau reported to the Convention on "a machine of war presented by the citizen Bonnemain" which was to destroy enemies of the Republic, but it was deemed imprudent to give a public description of it. Bonnemain made great claims for it and preliminary investigation had been favorable. The Convention decreed that he be allowed 6,000 livres for completing this weapon "without delay."[30]

The most important of these secret weapons, or the one for which the most extravagant claims were made, was one reported to the Council of the Five Hundred on 11 prairial an VII (May 30, 1799)—a weapon which would "exterminate an army of a hundred thousand men in a moment."[31] It would appear that its inventor had discovered the atomic bomb. History, however, has no record of any weapon of such devastating effects at that time. These extravagant claims are of interest, nonetheless, as evidence that experimentation and invention were in the air.

The Revolutionary governments on their part were eager to support inventors when convinced that they were working on something that might lead to success. One of the many aided was a certain Le Bon, who in April, 1792, was recommended for

[26] *Moniteur universel*, XII, 133.
[27] *Ibid.*, XIII, 395-96; Tuetey, *Répertoire générale*, V, 335.
[28] Tuetey, *Répertoire générale*, IV, 177.
[29] *Ibid.*, VII, 258.
[30] *Moniteur universel*, XVII, 714.
[31] *Ibid.*, XXIX, 689. This last claim was read and ignored by the council.

a grant of 2,000 livres for working on the improvement of firearms. Le Bon had previously submitted the prize-winning essay in a contest of the School of Bridges and Highways.[32]

Another award of 5,000 livres was recommended to the Legislative Assembly in May and June, 1792, as recompense to the cannoneer Nanin for an ingenious invention for the rapid unspiking of a cannon and for certain other devices.[33]

The leading attempts at bettering gunpowder have already been described in the chapter on chemistry. Prior to the 1700's little attention was given to its improvement. Such developments as had occurred in firearms had been in mechanical construction. Increasingly toward the close of the eighteenth century, however, new substances were tried out in the place of saltpeter, with interesting results. In this development the names of Lémery, Bayen, Fourcroy, Vauquelin, Lavoisier, and Berthollet stand out.[34] Bayen, chief pharmacist in Louis XV's armies, made known the explosive properties of mercury fulminate; but he did not propose its use for military purposes, nor did anyone else at that time. Berthollet in 1788 recommended the substitution of chlorate of potash for saltpeter in cannon powder. He observed that when mixed with sulphur and charcoal, chlorate of potash became a powerful explosive. Lavoisier, too, suggested its substitution for gunpowder, but with the fatal explosion of October 27, 1788, when Le Tort, factory superintendent, and Mlle. Chevraud, who were in Lavoisier's party of spectators, were killed, all further experimentation was dropped.[35] During the 1790's Berthollet, Fourcroy, and Vauquelin experimented with fulminates, with results already described.

There were few notable inventions in military art and arms in the eighteenth century, yet there was much evidence of innovation and intellectual striving toward invention. The one major invention—the principle of the nation in arms—can be reckoned among the most significant developments of the century.

[32] Tuetey, *Répertoire générale*, VII, 264.
[33] *Ibid.*, 266.
[34] Figuier, *Merveilles*, III, 473-74; *Nouvelle biographie générale*, IV, 865-66; William Wellington Greener, *The Gun and Its Development, with Notes on Shooting* (2d ed., London, 1884), 107; Kiréevsky, *Législateurs chimistes;* Richard, "Savants et salpêtre," 231-36.
[35] Sidney J. French, *Torch & Crucible: The Life and Death of Antoine Lavoisier* (Princeton, 1941), 170-71.

Chapter XI

Medicine and Surgery

IN MEDICINE PROPER THE FRENCH WERE NOT OUTSTANDING in the eighteenth century. Their leading medical schools, the universities of Paris and Montpellier, were inferior to those of Leyden, Edinburgh, and Vienna. On the other hand, French surgery was unexcelled, and was rivaled only by that of Great Britain, where in the second half of the century a number of notable Scots set the pace. Petit, Mareschal, La Peyronie, Chopart, Baseilhac, Le Cat, Janin, Daviel, Puzos, Lebas, Desault, Bichat, Coysevox—these are names of remarkable French surgeons of the period, and they do not exhaust the list.

The physicians would no doubt be tremendously humiliated could they return to life and learn the verdict of historians that their own achievements were so outstripped by those of the surgeons. For in the eighteenth century the physicians lorded it over the surgeons, and not until the time of the Revolution were the two groups placed on an equal basis. The physician was university-trained, while few surgeons were. Moreover, medicine had made greater strides and had reached a higher eminence than surgery.

Why then did the surgeons so surpass the physicians? The answer lies only in part in the statement that they had farther to go. It rests much more in the willingness of the surgeons to experiment. It may also be partially attributable to the facts that the Academy of Surgery (1731) antedated the Academy of Medicine (1776) and that the organization and association of the surgeons were even better than those of the physicians, despite the lower status of the former.

Perhaps the most outstanding contribution to French medicine during the century was in the study of membranes made, interestingly enough, by a surgeon, François Xavier Bichat (1771-

1802). A native of the French Alpine village of Thoirette and son of a physician, young Bichat was sent to the nearby Collège de Nantua for his general education, and later became the pupil in turn of two celebrated surgeons, Marc Antoine Petit of Lyons and Pierre Joseph Desault of Paris, the latter regarded by historians as the most eminent French surgeon of the last three decades of the century. A brilliant report of a class lecture brought him to the attention of Desault, who, after a conversation with the young man, invited him into his home and at length into his practice and research. This high confidence did not go unrequited. A noted teacher and an indefatigable worker, he labored by day at his teaching, practice, and research, and devoted his nights, after Desault's death (1795), to editing the papers left by his great benefactor. In the period 1799-1802 he published three notable treatises setting forth the fruits of his own researches, of which the first, *Treatise on Membranes* (1799-1800), is today regarded as the foundation of the modern science of histology. His treatise was devoted to the study of the effect of disease upon various tissues of the body. Twenty-one specific tissues were studied and described with remarkable clearness. His untimely death caused by a brain concussion resulting from a fall on the staircase of the Hôtel-Dieu was a loss to France and the world. One American medical historian in praising him stated: "Probably there is the name of no other member of the profession on the pages of Medical History, who accomplished as much valuable scientific work in the brief period of eleven years, as was done by Bichat in the closing years of the eighteenth century."[1]

A second branch of the medical science was foreshadowed if not actually inaugurated by the physician Théophile de Bordeu (1722-1776). Born in the village of Iseste in the French Pyre-

---

[1] Nathan Smith Davis, *History of Medicine, with Code of Medical Ethics* (Chicago, 1903), 98-99. See also Albert H. Buck, *The Dawn of Modern Medicine: An Account of the Revival of the Science and the Art of Medicine Which Took Place in Western Europe during the Latter Half of the Eighteenth Century and the First Part of the Nineteenth* (New Haven, Conn., 1920), 162-68; D'Arcy Power and C. J. S. Thompson, *Chronologia Medica: A Handlist of Persons, Periods and Events in the History of Medicine* (New York, 1923), 182; Esmond R. Long, *A History of Pathology* (Baltimore, 1928), 127-28.

nees, he received his education in the Jesuit College of Pau and at the University of Montpellier. Graduating from the latter in 1744, he was for short periods teacher of anatomy at Pau and director of the baths in the Pyrenees, but in 1752 he moved to Paris and spent the remainder of his life there in private practice. Almost immediately on his arrival in Paris he published an "important and exhaustive memoir" entitled *Recherches anatomique sur la position des glandes, et sur leur action* (1752), calling attention to the bearing of the glands on the health of the body. In this and in his later treatise, *Recherches sur les maladies chroniques* (1775), he foreshadows the recent science of endocrinology. He held that all organs of the body have their secretions which move into the blood and when functioning well give the individual tone and health. Failure to function well, on the other hand, leads to ill health or weakening of the personality. In particular De Bordeu made studies on the effects of testes and ovaries upon individuals. He observed eunuchs, capons, and spayed animals— their tendency to obesity, their retiring nature, and so forth. Medical historians seem reluctant to credit him with actually originating the science of endocrinology, partly because his observations were not supplemented by experimentation, and partly because his books and ideas collected dust for some decades before the subject was revived.[2]

That branch of medicine known as psychiatry likewise had its origins in the eighteenth century. To the French physician Philippe Pinel (1745-1826) is commonly attributed the honor of its founding, because of his humanitarian treatment of the insane. Another great Frenchman from a rural hamlet, Pinel was born in Languedoc, received his medical training at the universities of Toulouse and Montpellier, and in 1793 was appointed chief physician at the Bicêtre in Paris, where, with government permission,

---

[2] Buck, *Dawn of Modern Medicine*, 159-62; Johann Hermann Baas, *Outlines of the History of Medicine and the Medical Profession*, tr. by H. E. Handerson (New York, 1910), 624-25; Iago Galdston, *Progress in Medicine: A Critical Review of the Last Hundred Years* (New York and London, 1940), 156-58; Fielding H. Garrison, *An Introduction to the History of Medicine* (3d ed., Philadelphia and London, 1924), 376-77; Arturo Castiglioni, *A History of Medicine*, tr. by E. B. Krumbhaar (New York, 1941), 586-87. Galdston goes farther than the others and recognizes De Bordeu as the "first theorist" of endocrinology.

he began his practice of psychiatry upon the insane imprisoned there. Certain of the patients experiencing delusions of grandeur were clearly paranoic. Couthon, crippled member of the committee of public safety, asked Pinel if he himself had not gone mad when he wished to turn loose such creatures. Yet Pinel suffered no harm from them, and thereupon began "the modern 'open-door' school of psychiatry." At least this is the accepted story.[3]

Actually these ideas did not originate with Pinel, nor was he the first to work with the insane with the idea of restoring sanity. Alexander Hunter, a Scottish physician, was engaged in this work in an asylum at York, England, in the late 1780's, and his work was known to the French government in early 1790, when an attempt was made through correspondence to learn more about it.[4] Likewise in Italy a daring physician named Vincenzo Chiarugi, appointed director of an asylum for the insane in 1788, began work with the inmates in the hope of curing them. In 1802 he was given a university chair to teach psychiatry. Another whose ideas ran along the same line was Joseph Daquin, a native of Chambéry, France, who in 1791 published a book entitled *Philosophie de la folie*, advocating the abolition of chains and of confinement in cells for the insane.[5] Nor were these the only pioneers in this field. They illustrate the fact that the idea was current at the time. Pinel was not the first to apply ideas of psychiatry to the insane; yet it was he who gave them impetus.

Certain writers have claimed that the French Protestant pastor of Montpellier, Jacques Antoine Rabaut-Pommier (1744-1820), anticipated the English physician Edward Jenner in the idea of using vaccination as a preventive against smallpox.[6] According to these claimants, Rabaut-Pommier observed that sheep and cattle

[3] Garrison, *History of Medicine*, 435-36; Baas, *History of Medicine*, 638-39; Paul Bru, *Histoire de Bicêtre (hospice-prison-asile), d'après des documents historiques* . . . (Paris, 1890), 454-58.

[4] *Procès-verbaux et rapports du comité de mendicité de la Constituante, 1790-1791*, ed. by Camille Bloch and Alexandre Tuetey (Paris, 1911), 19, 163-64; *Archives parlementaires*, XXII,

631; McCloy, *Government Assistance*, 228 n. 84.

[5] Castiglioni, *History of Medicine*, 633-34; Joseph Coiffier, *L'assistance publique dans la généralité de Riom (au XVIIIe siècle)* (Clermont-Ferrand, 1905), 169 n. 2.

[6] Lacroix, *XVIIIme siècle*, 57; *Nouvelle biographie générale*, XLI, 384.

were subject to a disease called *picote* (cowpox), and that shepherds or milkmaids contracting the disease were thenceforth immune to smallpox. Around 1784 he himself made some experiments and sent a verbal message about his ideas to Eward Jenner through two English travelers at Montpellier, James Ireland and a Dr. Pugh. Years afterward Ireland reported that he had communicated the message to Jenner. Hence Jenner, it is claimed, was indebted for his idea of vaccination to Rabaut-Pommier. On the other hand, Jenner made no mention of it, although in his published works he gave historical details of his idea and experiments; nor do Jenner's biographers or English or American medical historians refer to the story. For lack of sufficient evidence, therefore, the claim for Rabaut-Pommier, highly educated man though he was and member of a distinguished family, must be rejected.[7]

Another Frenchman in 1745 experimented with inoculation as a preventive against the cattle plague (rinderpest) and met with partial success. It is the method used today in preventing this dread cattle disease. Had De Courtivron met greater success or continued his experiments longer, France and not England would have the credit for this achievement. The experimenter in this case was, interestingly enough, a French aristocrat and soldier who turned scientist after being badly wounded in the War of the Austrian Succession. Working under the tutelage of the renowned Clairaut, he became a scientist of distinction in his own right, skilled in physics, astronomy, and mathematics, and making important contributions in mechanics and optics. The Academy of Sciences received him as member. His experiments in 1745 came during a severe outbreak of the cattle plague in France. To determine whether the disease was communicated by contagion or through the air, he placed over two healthy cows the hides of some cattle killed by the disease. The cows undergoing this experiment contracted the disease in such a mild form

[7] N. B. Camac (comp.), *Epoch-making Contributions to Medicine, Surgery and the Allied Sciences; Being Reprints of Those Communications Which First Conveyed Epoch-making Observations to the Scientific World, Together with Biographical Sketches of the Observers* (Philadelphia and London, 1909), 216-22, 236-37.

that De Courtivron decided that they did not have it. He was interested in the experiment and recorded it in a pamphlet on the epizootic, but unfortunately he did not realize the value of his work. In 1748 he repeated the experiment with the same conclusion.[8] In less than a decade three Englishmen, Dobson, Fleming, and Layard, as a result of experimentation became convinced that the disease could be prevented by inoculation, and the last two published pamphlets on the subject in 1755 and 1757 respectively.

There were no doubt developments in the treatment of other diseases. Numerous drugs ("remedies" they were called) were submitted to the government for approval and adoption, some by physicians, some by pharmacists, and some by sheer quacks. The greater number were rejected, but not a few were approved. The government paid the owners for their formulas, either in a lump sum or by a pension, and required in each instance that they publish the formula, and that two copies of it be filed, one with the secretary of state for the Maison du Roi.[9] The government even distributed gratis through the intendants and their subdelegates large quantities of these drugs to the needy.[10]

Among those pensioned for their remedies were Denis Claude Doulcet (1722-1782), physician at the Hôtel-Dieu of Paris, where he devised a means for combating puerperal fever which according to the hospital records had remarkable success.[11] Likewise, in 1786, a physician named Courcelles and two assistants were given 4,000 livres in pensions for an elixir they concocted for the pre-

[8] M. J. A. N. Caritat, Marquis de Condorcet, *Oeuvres de Condorcet*, ed. by A. Condorcet O'Connor and M. F. Arago (12 vols., Paris, 1847-1849), III, 187-95; Jean Jacques Paulet, *Recherches historiques et physiques sur les maladies épizootiques, avec les moyens d'y remédier dans tous les cas* (2 vols., Paris, 1775), I, 217-27; *ibid.*, II, 134-42; *Histoire de l'Académie, 1745*, pp. 3-4.

[9] Full details can be found in the royal ordinance of April 12, 1776, in A. R. J. Turgot, *Oeuvres de Turgot*, ed. by Eugène Daire (new ed., 2 vols., Paris, 1844), II, 473-74. All pharmacists were required to inscribe these remedies in a register for their own use.

[10] McCloy, *Government Assistance*, 110-14.

[11] Doulcet and his family were pensioned, the government published his book on the subject, and an award was given the midwife assisting him. *Ibid.*, 164, 330, 332; *Documents des hôpitaux de Paris*, II, 127, 130, 133-34. His biographer in *Nouvelle biographie générale*, XVI, 693, says that he used ipecac and *sal de duobus* (potassium sulphate) and doubts their value.

vention of disease to women in childbirth.[12] A physician named Keyser and later his sister were pensioned because of an antivenereal remedy that came to be widely used with alleged success in the army hospital in the last decades of the Old Regime.[13] Other venereal drugs, by Mittié (1727-1795), physician to King Stanislas of Lorraine, and by Laffecteur, were praised by contemporaries but failed to win government support. Mittié used a treatment without mercury; Laffecteur, a rub in which likewise was no mercury. Mittié was a distinguished physician; of Laffecteur little is known save that he lived in Paris and in 1778 obtained the approval of the Royal Society of Medicine for his rub, which he claimed had healed numerous persons in a desperate condition.[14]

Louis XIV bought and published a formula, devised or at least possessed by a surgeon in Berry named Brossard, using agaric of oak for coagulation of the blood. Likewise Louis XVI bought from a widow named Nousset in the Swiss mountains a formula which was reputed to be an infallible cure for tapeworms.[15]

This list of remedies bought and given government encouragement could well be extended, but to little purpose. They added little if anything to knowledge of the body; they added little in arriving at permanent antidotes and cures. To be sure, there was much experimentation, but it was largely blind. It was necessary that chemistry and microscopy advance before the way would be opened to further medical development. As it was, the greater number of eighteenth-century physicians did not recognize the difference between syphilis and gonorrhea, or between typhus and typhoid. The germ nature of disease was unknown. Several

---

[12] *Archives parlementaires*, XIV, 71, 96. The two assistants, Jean Baptiste Chevillon and Joseph Desportes, were each given a pension of 1,000 livres.

[13] McCloy, *Government Assistance*, 33, 165-66.

[14] Tuetey, *Assistance publique*, IV, 531-32; Augustin Cabanès, *Chirurgiens et blessés à travers l'histoire, des origines à la Croix-Rouge* (Paris, 1918), 332 n. 4; Pierre Boyveau-Laffecteur, *Adresse à l'Assemblée nationale* (Paris, 1791), 1-4; *Nouvelle biographie générale*, XXXV, 707-708; *Inventaire-sommaire des archives départementales antérieures à 1790. Côte d'Or*, ed. by Claude Rassignol and Joseph Garnier (Paris, 1864), xix, C 363 (hereafter *Archives de la Côte d'Or*); *Archives de la Gironde*, C 3670; *Archives de l'Ille-et-Vilaine*, C 91; Paris *Mercure de France*, April 2, 1791. Cabanès, with reason, ridicules the idea of treating syphilis with vegetal concoctions.

[15] Lacroix, *XVIIIme siècle*, 38; Legras, *Notice historique*, 47-48.

French physicians and surgeons arrived at it by theory, but until the microscope was sufficiently developed, there was no way to prove it, or at least no way to convince the world of it.[16]

Surgical progress was marked by new methods and instruments. Developments in lithotomy were made by Jean Baseilhac (1703-1781), Antoine Louis (1723-1792), Claude Nicolas Le Cat (1700-1768), and others. Baseilhac was known commonly as Frère Côme, a name that he received on entering the order of Feuillants after he had received surgical training at the Hôtel-Dieu of Lyons and the Hôtel-Dieu of Paris. Allowed by his order to continue his service as surgeon, he gave special attention to operation for bladder stones. At length he developed a new method entitled the "hidden stone operation" *(lithotome caché)*, because it was made through the rectum. The idea was first suggested to him by the lateral operation made by Frère Jacques de Beaulieu, a wandering surgeon of considerable reputation, but the method at which he arrived seemed simpler and safer to him. Nevertheless for two years he practiced upon cadavers before attempting the operation on a living person. It met with success, and Baseilhac came to be greatly sought after, so much so that in 1753 he set up a small hospital in Paris where he treated the poor gratuitously. Certain of his contemporaries severely criticized his method, but he did not waver; rather, he published replies to them. He left three separate treatises describing his method.[17]

The contribution by Marc Antoine Louis was termed the "double lithotomy," performed on women. Louis was a leading surgeon in Paris, holding many positions of distinction, among them the permanent secretaryship of the Academy of Surgery and the inspectorship of military hospitals. In addition to this new form of operation, he invented several surgical instruments— some curved scissors and a surgical knife for amputations. He

[16] Cabanès, *Chirurgiens et blessés*, 332; McCloy, *Government Assistance*, 146 n. 46, 162; *Nouvelle biographie générale*, XXII, 943-44.

[17] *Mémoires de l'Institut*, II, 341-63; *Grande encyclopédie*, V, 580; Castiglioni, *History of Medicine*, 624. Castiglioni says that Baseilhac also operated above the pelvic bone. Thousands underwent operation at his hands. He also invented several surgical instruments, and devised an oblique method for extracting cataracts. Whether among his instruments he invented certain pincers that he described for the breaking of bladder stones is not clear.

also introduced medical jurisprudence in France, a procedure created earlier in the century by Michael Bernhard Valentine of Germany.[18]

One of the most severe opponents of Frère Côme was Claude Nicolas Le Cat (1700-1768), celebrated surgeon of Rouen, who himself followed the method of operation for stones accredited to the Englishman Cheselden. Not content to oppose Frère Côme in writing, Le Cat went to Paris and demonstrated before the Academy of Surgery the superiority of his technique. It appears that he was original in advocating that in operations for the stone the outer incision should be much wider than the inner, and that in healing the wound should be reduced "in size from without inward." Le Cat was also the inventor of two instruments for extracting bladder stones, one of them a small knife attached to a catheter and inserted through the urethra. He also claimed to be the inventor of a method of operation for lachrymal fistula. Le Cat was a native of Blérancourt, Picardy, but spent his professional career in Rouen, where he was chief surgeon at the Hôtel-Dieu, and where he founded an academy of arts and sciences. At the outset of his career he won several annual prizes offered by the Academy of Surgery at Paris and rejected an offer to establish himself there. He was finally given a pension and letters of nobility by the French government in recognition of his services.[19]

Operation for mastoiditis was originated by Jean Louis Petit (1674-1750), commonly regarded as "the most eminent surgeon of the first half of the century and the greatest contributor to his subject in France since the time of Paré." Born and trained in Paris, Petit was a precocious student and at sixteen was called upon to assist in the instruction of his fellow students, which he did satisfactorily. For some years during the 1690's he served as military surgeon with the army, but later returned to Paris where he was actively engaged in surgery for a half century. He was the founder of the Academy of Surgery (1731) and a member of the Academy of Sciences. He invented the screw tourniquet, first described softening of the bones and the development

---

[18] *Grande encyclopédie*, XXII, 668; Garrison, *History of Medicine*, 384; Castiglioni, *History of Medicine*, 637.

[19] Baas, *History of Medicine*, 663; *Nouvelle biographie générale*, XXX, 179-82.

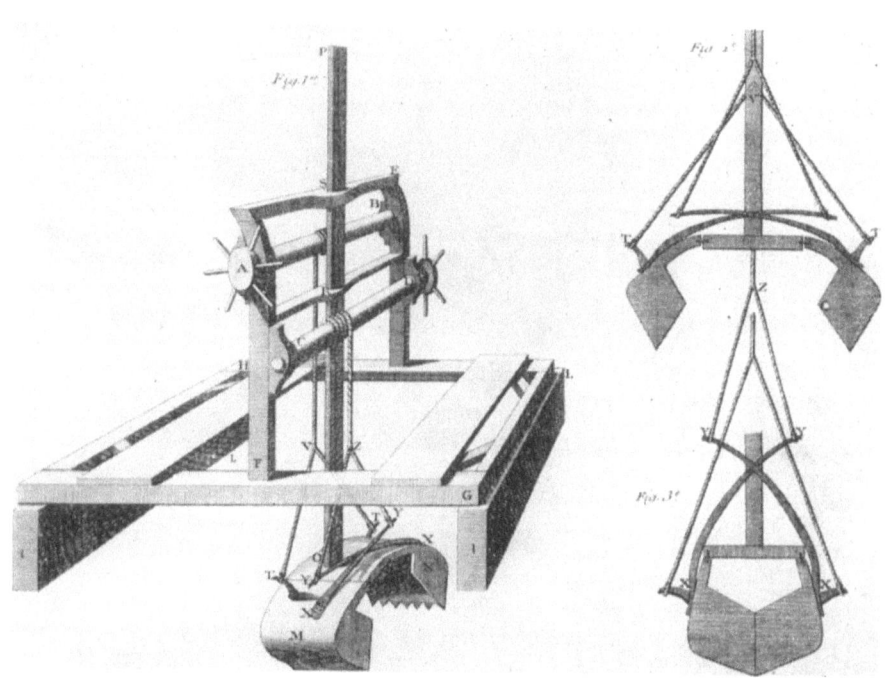

A DREDGE FOR CLEANING HARBORS

Gouffé invented this device in 1703.

(Illustration from *Machines approuvées par l'Académie*, II, 156, courtesy of the Ohio State University Library)

of blood clots in injured arteries, possibly first recognized that decayed or ill-formed teeth could be the cause of disease and called for their extraction, founded gall-bladder surgery, and made contributions to the modes of amputation and of operation for hernia. It is no wonder that the French regard him so highly. It is said that "he possessed a profound knowledge of the anatomy of almost every region of the body" and that he "seemed to possess, on the spur of the moment, an intuitive knowledge of what he should do in any situation of affairs that might suddenly develop in the course of an operation." While he wrote little, he did publish one remarkable three-volume work in 1774 (later republished in 1790) entitled *Traité des maladies chirurgicales et des opérations qui leur conviennent*. In it he describes some of his own experiences, such as the mastoid operation and extraction of teeth for ailments that had long baffled other physicians and surgeons.[20]

The first recorded case of operation for appendicitis was that by Mestivier in 1759, but it terminated unsuccessfully. Mestivier made an autopsy and wrote an account of the case, but the surgical profession was unimpressed and continued to neglect so promising a field.[21]

François Chopart (1743-1795), pupil and close friend of the great Desault and after 1771 professor of surgery at Paris, "was a pioneer in the surgery of the urinary tract," and developed a mode of amputating the foot (1792), which still bears his name.[22]

Nicolas Puzos (1686-1753), long a military surgeon before returning to Paris in a civilian capacity, acquired a great reputation at child delivery (then the province of surgeons rather than physicians) and arrived at a method of saving both mother and child in certain difficult cases by perforating the fetal membranes and inducing labor.[23]

---

[20] Buck, *Dawn of Modern Madicine*, 215-18 (where there are some excerpts from the *Traité*); Cabanès, *Chirurgiens et blessés*, 216-17; *Grande encyclopédie*, XXVI, 526; W. R. Bett (ed.), *A Short History of Some Common Diseases* (London, 1934), 150-51.

[21] Garrison, *History of Medicine*, 352.

[22] Buck, *Dawn of Modern Medicine*, 219-20. Buck quotes a description of this amputation from Thomas L. Stedman's *Medical Dictionary*: "Disarticulation at the metatarsal joint, leaving only the soft parts of the sole of the foot to cover the stump." In short, the fore part of the foot is cut off and the flesh turned under to form a pad.

[23] A member of the Academy of Surgery from its origin, Puzos succeeded

Jean Pierre David, a surgeon of Rouen, in 1779 set forth a graphic account of spinal deformity resulting from decayed teeth, and in 1782, contemporaneously with Pott in England, described necrosis of the bone.[24]

The first attempt to catheterize the Eustachian tubes from the mouth was made unsuccessfully in 1724 by Guyot, postmaster at Versailles. In 1741 this operation was performed with success by the Englishman Cleland.[25]

Plastic surgery in Europe (1800) has been credited to a French surgeon named Lucas, who served with the army in India and there witnessed such operations. While in India he had seen Tippoo Sahib, sultan of Mysore, cut off a man's nose and a native physician restore it with skin taken from the forehead. Lucas called his art "rhinoplasty."[26]

In ophthalmology the surgeon Dominique Anel (1679-1730), born at Toulouse and educated there and at Montpellier, made a remarkable surgical contribution by healing a lachrymal fistula (1710) by draining it, first with a wild boar's bristle and later by a small cannula or syringe which he invented. This syringe and a probe that he devised still bear his name. He is said, moreover, to have originated the mode of treating aneurysms prior to Hunter, who is commonly given the credit.[27]

Perhaps the greatest glory of French surgery of the eighteenth century, however, was the operation for cataract by Jacques Daviel (1750). Actually it was not the first time that this operation was performed, inasmuch as it was employed by Arabic physicians of the late Middle Ages. Later, in the sixteenth century, a German surgeon named Georg Bartisch wrote a treatise on the structure of the eye and the means of operation for the extraction of cataracts. Several French oculists of the early

---

Petit as royal censor of works on surgery and in 1754 was created a noble. *Nouvelle biographie générale*, XLI, 223.

[24] Garrison, *History of Medicine*, 349.

[25] *Ibid.*, 360; Baas, *History of Medicine*, 666. Baas credits the founding of otology to the French physician Duverney (1648-1730), who taught anatomy at Paris and published in 1683 a small work entitled *Traité de l'organe de l'ouie, contenant la structure, les usages et les maladies de toutes les parties de l'oreille*. It was republished in 1718 and 1731.

[26] Rambaud, "Sciences," 145.

[27] Baas, *History of Medicine*, 666; Roswell Park, *An Epitome of the History of Medicine* (2d ed., Philadelphia, 1899), 215; *Nouvelle biographie générale*, II, 631-32.

eighteenth century studied the nature of cataracts and even experimented with operation for them. Thus, in 1706, Antoine Maître-Jean (1650?-1730) published a treatise in which he pointed out the true nature of the cataract, namely the hardening and clouding of the crystalline lens. About the same time, Michel Brisseau (1677-1743) of Tournai, working independently, came to the same conclusions. In 1709 he wrote on glaucoma. In the 1730's and 1740's François and Etienne du Petit, father and son, worked at removal of the cataract with some success. Daviel's priority in operation for removal of the crystalline lens for cataract is thus open to question, and it has been heatedly debated. Regardless of whether he was the first in point of time to meet success at this operation, it was he who convinced Europe and the world of its practicality and popularized its use.[28]

Jacques Daviel (1696-1762) was a native of the small town of La Barre, Normandy. His surgical training was received from an uncle in Rouen and a surgeon named Bourdon at the Hôtel-Dieu in Paris. He was one of several young surgeons from the staff of this hospital sent by the government in 1720 to combat the terrible epidemic of bubonic plague in Provence.[29] The mortality for surgeons was high, but Daviel came through safely, rendered splendid service, and was honored by the city of Marseilles and the national government. For a time he remained at Marseilles as staff surgeon for a galley and as professor of anatomy and surgery. From 1728 he specialized on study of the eyes and acquired such a reputation that he was invited on missions to several Italian cities, and was made a member of the Academy of Toulouse and an associate member of the Academy of Surgery at Paris. In 1746 he returned to Paris and obtained permission to operate at the Hôtel des Invalides. In 1747 he operated on 206 persons for cataract, with success in 182 instances. The king of Spain, Ferdinand VI, tried to prevail on Daviel to come to Spain, but he declined. Daviel had already, in 1750, been appointed royal oculist in France. Several European academies made him

[28] Castiglioni, *History of Medicine*, 389-90, 484, 631-32. Denis Diderot, *Oeuvres complètes de Diderot*, ed. by J. Assézat and M. Tourneux (20 vols., Paris, 1875-1877), I, 280, tells of the Prussian oculist Hilmer operating for cataract in 1749.

[29] For an account of this epidemic see McCloy, *Government Assistance*, Ch. VII.

member or associate member. But he lived to enjoy his European glory only a few years. His health gave way, and after several unsuccessful attempts to regain it he died at Geneva in 1762, the leading French, if not also European, oculist of the century.[30]

In dental surgery enormous development was made during the century, the greatest impetus coming from the Frenchman Pierre Fauchard (1690-1761), commonly described as "the founder of modern dentistry." Born in Brittany, Fauchard practiced dentistry in Paris and in 1728 published a notable work entitled *Le chururgien dentiste (The Surgeon Dentist)*, which was revised and republished in 1746 and 1786. In 1733 it was translated and published in German. A work of more than 800 pages with many illustrations, it gave a complete account of dental science as practiced at that time, and "was highly commended by the leading medical authorities of the day." Not only was Fauchard a pioneer in giving such an account of his profession, but he has been credited as the first to describe pyorrhea (1746) and the first to use orthodontia in treatment of malformed teeth.[31] Some of the misplaced teeth he could straighten in a few days with his file and pelican; others required months, and for them he used metal plates with threads harnessed to the unruly teeth. He told of making plates with full sets of teeth for both upper and lower jaws, which satisfied their users, but it is not clear whether this was an innovation. He had more difficulty with the upper plates, which he contrived to hold in place by means of springs. He suggested the making of porcelain teeth, but they had been proposed and possibly even made as early as 1710 by Guillemeau, who devised a formula consisting of white wax and gum elemi with powdered white mastic, coral, and pearl. He claimed that

---

[30] *Nouvelle biographie générale*, XIII, 242-43; Garrison, *History of Medicine*, 358. According to Garrison, Daviel by 1750 had performed 434 operations for cataract, with failure in only 50.

[31] James A. Taylor, *History of Dentistry: A Practical Treatise for the Use of Dental Students and Practitioners* (Philadelphia and New York, 1922), 51, 55, 58; Garrison, *History of Medicine*, 356. Fauchard is known also for his famous recommendation of rinsing the mouth morning and evening with fresh urine as a remedy against toothache. This was a remedy often prescribed by physicians and surgeons of the time for other ailments. Fauchard claimed that it would not only stop toothache after a few days, but that according to the report of his patients it cured other troubles too. "Most people have some difficulty at the beginning to accustom themselves to it," he wrote, "but what will one not do for relief and for health?" Quoted by Taylor, *History of Dentistry*, 65-66.

such teeth would not turn yellow. Later in the century the surgeon-dentist Nicolas Dubois de Chémant did make a set of porcelain teeth for a porcelain manufacturer of Paris named Guerhard, who was dissatisfied with his teeth of hippopotamus ivory (then commonly in use) because they stank. Dubois de Chémant in 1788 published an account of his experiments and in 1789, on recommendation from the Academy of Sciences and the Faculty and Academy of Medicine in Paris, was granted a patent by the government. In 1790 he offered his artificial teeth to the general public with a lifetime guarantee that they would not lose in beauty or service, nor need replacement every year as hippopotamus bone required, nor emit an odor.[32]

In addition to new fields and modes of operation, French surgery made many contributions of a mechanical nature. Some have been described in passing. Among the most significant was the screw tourniquet (1718) by Jean Louis Petit, a device still used in the twentieth century for checking the loss of blood in the limbs, especially in amputation of the legs. It consists of two small boards or pieces of metal made to receive a large vertical screw with thumbpiece, the whole tightly bound to the leg by a bandage. The device is simple but remarkably effective.[33] It had advantages in that the surgeon could apply it and keep it on without assistance, and in that it applied pressure only to the artery or arteries involved.

To attempt an account of all the many instruments would be both wearisome and impossible. It may not be amiss, however, to name a few of the more important, which included a cleverly devised spring forceps for compressing an artery in an operation, by Desault (1787);[34] a number of obstetrical forceps, by Jacques Mesnard, André Levret, and Jean Louis Beaudelocque; a probe for bladder stones, by Henri François Ledran; a nasal probe or dossil, by Jean Louis Belloc, that might be inserted dry or sat-

[32] Taylor, *History of Dentistry*, 46, 62-64; *Journal de Paris*, January 13, 1790, supplement, iv. According to Castiglioni, *History of Medicine*, 881, Philipp Pfaff, dentist to Frederick the Great, published in 1756 "a more remarkable book" on dentistry than that by Fauchard.

[33] Cabanès, *Chirurgiens et blessés*, 215, 217; Charles J. S. Thompson, *The History and Evolution of Surgical Instruments* (New York, 1942), 85-86. Both works give illustrations of the device.

[34] Thompson, *Surgical Instruments*, 68; Buck, *Dawn of Modern Medicine*, 224. Thompson gives an illustration.

urated with a liquid; a knife for operation on cataracts and a bow saw with detachable blade, by Georges de la Faye; and a series of surgical knives and other instruments by Jean Jacques Perret, who though a surgeon and skilled anatomist refused to practice his profession but gave his attention to designing and manufacturing surgical instruments.[35]

A number of bandages were designed. A bandage or satchel for holding hernias in place was invented by the engineer Abeille (about 1742), who himself suffered from a hernia and contrived this means of support. In 1749 an elastic bandage for this same ailment was devised by Neilson, a Scottish surgeon practicing in Paris.[36] Later in the century Desault originated a skillful bandage, still known by his name, for holding the arm in correct position in cases of broken collarbone. The essential feature seems to have been the placing of a cushion on the chest for holding the arm at a proper angle from the body, so as not to exert improper pull.[37] Other bandages, including one for hernia, were invented around 1792 by Oudet, member of the Royal College of Surgery and the Society of Inventions and Discoveries. He claimed that his hernia bandage, which was for both sexes and all ages, allowed greater freedom of movement than those in use. He also invented a pessary, a urinoir, and an artificial leg ingeniously made to permit movement in any manner like a natural leg. Because of these inventions he was recommended to the king in 1792 by the Bureau of Consultation of the Arts and Trades for an award of 2,000 livres.[38]

Oudet, however, was not the first to make artificial limbs. François de la Noue (1531-1591), Breton nobleman and Huguenot soldier, used an iron arm after 1570, and in the late 1600's flexible artificial hands were made for a Swedish warrior by Jean Truchet and Du Quet of Paris. Pierre Joseph Laurent (1715-1773), a French engineer, made a clever mechanical arm for the Duc de

[35] Baas, *History of Medicine*, 664-65; Thompson, *Surgical Instruments*, 22, 30, 32; Power, *Chronologia Medica*, 149, 151, 171; *Nouvelle biographie générale*, V, 279-80; *ibid.*, XXX, 266; *ibid.*, XXXIX, 641.

[36] Both bandages are described and illustrated in *Machines approuvées par l'Académie*, VII, 191-200.

[37] See the account by Buck, *Dawn of Modern Medicine*, 223-24. It was Bichat's description of this bandage that brought him to the attention of Desault.

[38] *Moniteur universel*, XIII, 112; Tuetey, *Répertoire générale*, VII, 263.

la Vrillière. Mercier in his *Tableau de Paris* told of a certain Périer in Paris on the eve of the Revolution making artificial legs on which patients could even climb and descend stairs. The knees and ankles, controlled by steel plates, could be made to bend at will. Similarly he tells of artificial arms and hands, movable at will and fitted to the individual, fashioned by a certain Laurent, possibly a son of the Laurent mentioned above.[39]

Not without interest or significance was the creation of artificial anatomy by the skilled Mademoiselle Biheron (1719-1786), a friend of Diderot. A native of Paris and the daughter of a surgeon, she studied anatomy and set out to make wax representations of the different parts of the human body. To this enterprise she devoted forty-seven years, and her masterpiece was the body of a woman which she could open or dismantle. Every Wednesday she put her exhibits on display for the public, charging three livres a person. On the occasion of the visit of King Gustavus III to Paris, she lectured and exhibited her figures before the Academy of Sciences. In time the collection was bought by Catherine the Great of Russia through her ambassador. Curiously enough, it is reported that all the physicians in Paris, save Jussieu and Villoison, were hostile to her; nor did she have more supporters among the physicians and surgeons of London. She modeled her anatomical parts after cadavers that she kept in a glass cabinet in her garden. So lifelike were the representations that the Scottish surgeon Sir John Pringle is said to have remarked that all they lacked in being natural was the stink.[40]

To another brilliant woman, Madame Angelique Marguerite Leboursier de Coudray (1712-1789), Paris midwife, was also due a clever mechanical contribution to surgical science. This was the artificial body of a woman, contrived to teach midwifery by ocular means. A dummy infant was placed in the womb in every conceivable position and the teacher demonstrated to her observers how in each instance the delivery should be made. This

---

[39] Mercier, *Tableau de Paris*, VIII, 123-25; Fontenelle, *Oeuvres*, VI, 391-92; James Westfall Thompson, *A History of Historical Writing* (2 vols., New York, 1942), I, 554-55.

[40] Stéphanie Félicité Ducrest de Saint Aubin, Comtesse de Genlis, *Mémoires inédits de Madame la Comtesse de Genlis, sur le dix-huitième siècle et la Révolution française, depuis 1756 jusqu'à nos jours* (2d ed., 10 vols., Paris, 1825), I, 338-39; *Nouvelle biographie générale*, VI, 65; Power, *Chronologia Medica*, 162.

visual instruction was extremely effective, and for twenty-five years (1764-1789) this remarkable woman, who had obtained her training at the Hôtel-Dieu of Paris, toured the French provinces on requests of intendants and city officials, giving her two months' course of training to large groups of students. One of the greatest needs of eighteenth-century France was a larger force of trained midwives, especially in the rural regions. The surgeons performed the service of delivery for such patients as cared to receive and pay for skilled treatment, but the overwhelming majority of French women were too prudish or niggardly or poor to permit a male practitioner to attend them. The result was a terrible mortality rate in cases of childbirth, both of mothers and children. Keenly conscious of the need of increasing the population, the French government in the last three decades before the Revolution gave special attention and made no inconsiderable outlays to increase the deplorably inadequate medical services in rural regions, above all in the matters of midwifery and child welfare. Madame de Coudray was well paid, often receiving 300 livres or more for her course of lectures (when a Sister of Charity at nursing might receive no more than 150 livres a year). According to an entry in the famous *Livre rouge* of 1798, she was paid by the government a salary of 4,000 livres during the first six months of 1783. She remained on the government payroll until her death in 1789, and it cannot be doubted that she rendered a valuable service to her country. By 1777 she had taught more than 4,000 pupils, 400 at Dijon alone. Over and over again her name appears in the correspondence of the intendants of provinces with the controller general of finances, and few were the provinces where she did not lecture. She was also the author of a popular book on midwifery (1759), which went through six editions.[41]

In Britain a similar service was rendered by the Scottish surgeon William Smellie (1697-1763), who after early practice in

[41] McCloy, *Government Assistance*, 172-73; A. Dupuy, "Les épidémies en Bretagne au XVIIIe siècle," in *Annales de Bretagne*, III (1887-1888), 189-92, 197-200, 203; Power, *Chronologia Medica*, 156; *Archives de l'Ille-et-Vilaine*, C 1326; *Archives des Bouches-du-Rhône*, C 83 f374, 1021; *Archives de la Gironde*, C 3302. Power reproduces a drawing of Madame de Coudray.

his native Lanark attended some lectures on midwifery in Paris and thenceforth for twenty years practiced in London. Working as a male midwife, he "introduced an improved technique and trained more than nine hundred men, as well as many women, in his new methods." In his teaching he, too, made use of a dummy to simulate the undelivered infant. His work was accomplished in the face of much opposition, both from the midwives whose economic interest was at stake and also from a part of the public.[42] Smellie trained a smaller number of midwives than did Madame de Coudray (whom he anticipated), but his service was possibly greater inasmuch as the male midwives of modern times have as a whole been better trained than the female. His break with the past in medical tradition was therefore more radical than that of Madame de Coudray.

Aids for afflicted persons and for the treatment of the sick were invented. Among them may be mentioned several forms of ear trumpet for the deaf, invented by Du Guet and approved by the Academy of Sciences in 1705. These acoustic horns varied in size and shape. Invariably one end was to be placed in the ear. One of the instruments, shaped like a crescent, fitted over the head with a tube for each ear. Some were so small that they could be hidden under the wig or hair; others had to be held in the hand like a small horn, with orifice to the ear. One instrument was fitted to a chair, and a deaf person had only to sit in it to hear amplified sounds. The devices were all of some metal—copper, silver, but chiefly tin. Their essential feature was a parabola designed to catch sound waves and direct them to the ear.[43]

Jean Truchet (1657-1729), a Carmelite monk who went by the name of Père Sebastien, was also the inventor of an ear trumpet. The son of a merchant at Lyons, he early entered the Carmelite order and was sent to Paris, where oddly enough for a monk he devoted himself to the study of geometry and mechanics. When two watches designed to strike the hour *(montres à repetition)*,

---

[42] Richard H. Shryock, *The Development of Modern Medicine: An Interpretation of the Social and Scientific Factors Involved* (New York, 1947), 92; *D.N.B.*, XVIII, 399-400. This William Smellie was not the great naturalist and first editor of the *Encyclopaedia Britannica*, whom he anticipated slightly in age.

[43] *Machines approuvées par l'Académie*, II, 119-30.

gifts from Charles II to Louis XIV, became out of order, they were sent to a jeweler for repair; but the jeweler could not open them and called in Truchet, who succeeded. This brought him to the attention of the king, or at least of Colbert, who gave him a pension of 600 livres. Later Truchet had an important part in supplying the fountains of Versailles with water, and he was consulted on the construction of all canals in his day. Besides his ear trumpet, he invented a "devil" for transplanting large trees without injury.[44]

A new form of hospital bed, designed to facilitate treatment of the sick or injured, was invented by the celebrated textile designer, Philippe de Lasalle (1723-1804). In the closing years of the Old Regime great attention was given to the insanitary conditions of the hospitals, especially in the matters of ventilation and beds, and attempts were made to better them. In this undertaking the government had a leading part. At Lyons in particular, where Lasalle lived and maintained a manufacturing enterprise, there was great interest in the movement. A new Hôtel-Dieu with a magnificent dome and careful attention given to aeration, considered as a model hospital building in France, was erected shortly before the Revolution. As for Lasalle, he was twice pensioned for his inventions, first in 1753 with 600 livres, and later during the ministry of Turgot with 6,000 livres; and he was awarded the cordon of Saint-Michel. Few eighteenth-century inventors so enjoyed the smiles of government favor.[45]

Other inventions auxiliary to surgery included a hammock carriage for transporting wounded soldiers from the battlefield, invented in the 1790's by Saint-Sauveur, and "flying ambulances" by Larrey and Percy near the turn of the century for giving first-aid treatment to the wounded on the battlefield. Since the late 1500's litters had been in use for carrying wounded soldiers to hospitals, and in the reign of Louis XV carriages for this purpose came into vogue. The surgeon De la Faye (c. 1701-1781) devised an ingenious means of transporting soldiers whose legs

[44] *Nouvelle biographie générale*, XLV, 677-78; Fontenelle, *Oeuvres*, VI, 384-400.
[45] *Nouvelle biographie générale*, XXIX, 726. De Lassone, physician to the queen, was active in getting hospitals to use iron beds, which were advantageous in combating vermin. Condorcet, *Oeuvres* (O'Connor ed.), III, 304.

were broken by shot, and shortly afterward another army surgeon named Ravaton conceived of having brigades of surgeons and assistants render first-aid treatment on the field while the battle was yet in progress. Ravaton besought the French government without success to put his idea into operation.

During the French Revolution Ravaton's idea was applied by two celebrated military surgeons, Jean Dominique Larrey (1766-1842) and Pierre François Percy (1754-1825). It was Larrey who originated the service, in late 1792 in a Rhineland battle between the French and the Prussians. Three surgeons with an attendant *(infirmier)*, all mounted on horses with satchels holding their instruments and equipment, went on the battlefield. Such was their service that they were commended by the government representatives-in-mission present with the army of General Custine in January, 1793.[46]

Larrey, the creator of this service, had been called "the most illustrious of the military physicians of the epoch" (of the Revolution and Empire). Born in the little town of Beaudéan in the French Pyrenees, he was left an orphan at thirteen and received his surgical training from a kindly uncle near Toulouse and from Desault at the Hôtel-Dieu in Paris. For a brief period he was an assistant surgeon on a French war vessel, but the outbreak of the Revolution found him at the Hôtel-Dieu in Paris. In 1792 he volunteered as a military surgeon and in the wars of the next twenty-three years served in several theaters—the Rhineland, Italy, Egypt, Spain, Russia, and finally at Waterloo, where he was wounded and captured. Napoleon bequeathed him the large sum of 100,000 francs and called him "the most virtuous man I have ever known." At that time he was surgeon-in-chief of the French army. After his death the French erected a statue of him before their great military hospital of Val-de-Grâce. Besides the "flying ambulances," Larrey originated a treatment of wounds by irrigation—letting a constant stream of water flow on the wound for the removal of inflammation, reduction of suffering, and rebuilding of tissues. Sometimes irrigation made amputation unnecessary.[47]

[46] Cabanès, *Chirurgiens et blessés*, 294, 374.

[47] Rambaud, "Sciences," 143-44; Buck, *Dawn of Modern Medicine*, 243-49.

Several years after first-aid surgical attention on the battlefield was inaugurated, Percy, chief surgeon of the French army, in concert with Larrey enlarged and improved it, forming groups of surgeons for first-aid service and groups of stretcher-bearers, each with its distinctive uniform. Some of the stretcher-bearers went afoot, others in carriages. Each division had twelve light carriages for transportation of the wounded, some with two wheels, some with four. (Larrey had taken pains to design a new form of carriage with springs, whereby jolting of the injured was greatly reduced.) These "flying ambulance" corps were present on Napoleon's first Italian campaign (1796-1797). Not until the early 1800's, however, did they become a permanent feature of the French army.

Another small-town boy, Percy was born in Franche Comté, the son of a surgeon. He first took mathematical training preparatory to entering the artillery, but later changed to surgery and studied that subject first in Burgundy and later in Paris under the celebrated Antoine Louis. In 1782 he became an army surgeon and followed that career throughout the Revolutionary and Napoleonic periods, being present at Waterloo. He won a large number of prizes given by academies and was a member of the Academy of Sciences and the Academy of Medicine. Napoleon in 1809 made him a baron of the Empire. For his service to the wounded soldiers of the Allied armies in France in 1814, he was given a high Russian decoration.[48]

Likewise in order to introduce a more comfortable means of conveying wounded soldiers from the battlefield, Jacques Grasset de Saint-Sauveur (1757-1810) suggested his hammock-carriage. He freely waived in advance any claim to the rewards commonly given inventors. This gentleman, it seems, had obtained the idea in the Near East, where for a long period he had been in French diplomatic service. He had been born in Montreal, Canada, but had been taken early to Paris and there educated at the Collège de Sainte-Barbe. Thence he entered upon a diplomatic career. Whether the French army used his scheme is not indicated.[49]

[48] Cabanès, *Chirurgiens et blessés*, 388. *Nouvelle biographie générale*, XXXIX, 562-63, gives a biographical sketch of Percy.

[49] Jacques Woeuvre, "Le transport des blessés pendant la Révolution," in *Revue historique de Bordeaux et du département de la Gironde*, IX (1916),

All too commonly we think of the great advances in medicine and surgery as having been made within the past century, and it is of course correct to say that the greatest achievements have come in this period, thanks to the remarkable developments in microscopy and chemistry. Nevertheless, amazing developments had been made over a long period prior to it, as the achievements recorded in this chapter illustrate. In the eighteenth century the French did not sit idly by. Their physicians and more particularly their surgeons left a record of achievement of which they can well be proud. One American medical historian gives French surgeons credit for having led the field in the eighteenth century, and of having done more "than during any other equal period of time."[50] The Dutch medical historian Baas also credits the French surgeons with leading the field in the eighteenth century. Whether all medical historians may render the same verdict, however, is questionable. Certainly the French were excelled by none. And if a reason is to be given for their remarkable achievements, it probably can be laid to the readiness of their surgeons to experiment. They seemed to be constantly seeking some better way of doing things. They were proud of their profession and had a strong humanitarian spirit. They wanted to serve and create. The overwhelming majority of them were sons of surgeons. The dynastic tendency of both the French physicians and surgeons had the great value not only of fostering an *esprit de corps* but also of introducing the youth at a very early age to the techniques of the profession. As a rule the French surgeons received their training quite early, and it was acquired largely by apprenticeship.

302; *Nouvelle biographie générale*, XXI, 709-10; *Biographie universelle*, XVIII, 327. Woeuvre says that Saint-Sauveur got his idea from an article and drawing by a lawyer named Cornu in *L'encyclopédie des voyages*.

[50] Davis, *History of Medicine*, 101. He says: "In France, more than in any other country during the eighteenth century, greater developments were made, both in surgical practice and in the relative rank of surgical practitioners, than during any other equal period of time." Baas, *History of Medicine*, 680, after discussing French surgery, remarks that "the French undoubtedly took first place in obstetrics also, the natural result of their possessing the first institutions in this art."

CHAPTER XII

Patents and Encouragement

NOT UNTIL 1791 DID THE FRENCH GOVERNMENT GRANT PATents to inventors; however, an imperfect system of government recognition of the rights of inventors did exist. Louis XIV and Colbert in chartering the Academy of Sciences in 1666 designated as one of its functions the investigation of claims to scientific or mechanical invention or discovery. This body in each case appointed a committee of three or more members considered best qualified to pass on the merits of the claim. The applicant was required to submit a detailed written description of the item invented, and with it drawings illustrating both the mechanism as a whole and its component parts. The committee had the power to demand, as in the case of Jouffroy, that the inventor bring a sample of his invention to Paris for demonstration, but this was seldom asked. Sometimes in the Academy's early days the committee reports were brief and sarcastic, but over the decades its investigations were painstaking.[1] In cases where the decision was favorable, the applicant was given a certificate of approval.[2] This in itself was a mark of honor, and it opened the way for the applicant to petition the government for monopolistic rights in exploiting the invention.

Certificates of approval were given not only for inventions of new articles or machines but also for improved models, and even for devices that though new in France were already in vogue in other countries. This practice set pitfalls for the historian examining the records of French inventions. Happily the applications for approval not infrequently give historical data of previous developments.

[1] Bertrand, *Académie des Sciences*, 147-48.

[2] Mention of one granted Gensonné in 1737 for a paper mill is in *Machines approuvées par l'Académie*, VII, 203.

Monopolistic rights were granted commonly for fifteen years. When the invention was impressive and augured a valuable contribution to the national welfare, the inventor often was pensioned and given financial aid by the government to develop his enterprise. This happened, for example, in the instances of Vaucanson, Brizout de Barneville, and the Montgolfiers. The government was eager to develop the nation's economic life, and to this end made heavy subsidies of 5,500,000 livres in the period 1740-1780, and loans without interest of 1,300,000 livres.[3]

Unfortunately the government had no uniform policy for all inventors. Some it aided financially; others it did not. More flagrantly inconsistent was its policy of granting monopolistic rights in many lines to court favorites, so that the greater number enjoying exclusive privileges were not inventors.

In some parts on the eve of the Revolution there was a belief that inventors deserved better treatment at the hands of the government. This opinion found expression in certain of the *cahiers* of 1789.[4] There resulted the patent law of January 7, 1791, passed by the National Assembly to render justice to inventors and to encourage invention. It established the right of ownership of every invention to its inventor (art. 1). Anyone bringing an invention to France, though already applied elsewhere, would be regarded as the inventor and would enjoy the rights of such (art. 3). To insure an inventor in his rights, he would be issued a patent giving him temporary enjoyment of exploiting his invention (art. 7). Patents would be given for periods of five, ten, or fifteen years, according to the request of the inventor; but they might not be given for more than fifteen years without a governmental decree (art. 8). Rights granted for inventions already patented in other countries would not be permitted to extend beyond the time limit set in those countries (art. 9). All patents would be given in parchment, with the national seal affixed, and registered with the secretaries of the directories in all the departments in France (art. 10). Anyone might go to the depart-

[3] Bondois, "Industrie et commerce," 149.
[4] It is found in the *cahiers* of the bailliage and the third estate for the city of Havre de Grâce. *Cahiers de doléances du bailliage de Havre (sécondaire de Caudebac) pour les Etats Généraux de 1789*, ed. by E. Le Parquier (Epinal, 1929), 121, 198.

mental headquarters and consult the patents granted (art. 11). Persons infringing on the patent rights of others would be subject to confiscation of all such manufactured goods and instruments and to a fine of 6,000 livres, this last to be given to the needy in the district. For a second offense the fine was to be doubled (art. 12). On the other hand, if the owner of a patent brought charges against anyone for alleged violation of his rights and could not establish proof, he himself was to be subject to the fine that would have been imposed on the violator. The fine in this instance was to go to the accused rather than to the district's poor fund (art. 13). Every owner of a patent "would enjoy the privilege of setting up establishments throughout the whole extent of the kingdom" to apply the invention or discovery (art. 14).

On the expiration of each patent, its description was to be published so that anyone might exploit it (art. 15). Patents, moreover, might be voided under any of the following circumstances: (1) when the inventor was found to have given "an insufficient description"; (2) when the inventor used other (and secret) means not detailed in his description; (3) when the invention was already described in print in some European language; (4) when after two years the inventor had not put the invention or discovery into operation; (5) when after receiving a patent in France the inventor attempted to get the same idea patented in another country; or (6) when the holder of a patent violated in any way the patent and its obligations (art. 14).

The decree stipulated that holders of *lettres patentes* registered for inventions under the Old Regime would be granted continuance of their rights; but holders of mere *arrêts du conseil* or of *lettres patentes non verifiées* would be obliged to have their privileges converted into new patents (art. 18).[5]

The essence of this law was to treat the inventions as the peculiar property of their inventors and to give these inventors the right of exploiting them financially through monopolistic privilege

---

[5] *Moniteur universel*, VII, 4-5. Certain inventors suffered from this last provision, among them Argand, whose privilege for the manufacture of his lamps fell into public domain. Argand retired to England and later to Geneva. Figuier, *Merveilles*, IV, 26.

over a certain number of years. All inventors thus were placed on the same footing, and were assured of legal protection as long as they were faithful. The value of patents was all the greater because of the suppression of monopolistic privilege for favorites.

France was anticipated in the giving of patents only by Great Britain (1561) and the United States of America (1790). Continental countries other than France did not adopt the patent system until the nineteenth century: Prussia, 1815; Belgium, 1817; Austria, 1820; the members of the Prussian Zollverein, 1842.[6]

The body entrusted with the examination of claims to invention under the new French patent system was the Bureau of Consultation of the Arts and Trades, consisting of thirty members, half of them drawn from the Academy of Sciences, the remainder from other learned societies. All were to serve without payment, save for a chief secretary, six undersecretaries, and a servant, whose combined salaries were not to exceed 11,300 livres, and expenses of the Bureau were limited to 2,000 livres annually. It was created by a law of October 16, 1791.[7] After making examination of the merits of claims to invention, it was to report to the legislative branch of the government concerning them and their probable value to society, and to propose in each instance the reward, if any, to be given the inventor.[8] Its recommendations of monetary awards were in many instances not followed, due in part to financial straits of the government in that era of revolution and war, and in part to the hostility of certain administrators, as Roland, minister of the interior in 1792.[9] Among the able men serving in the Bureau of Consultation, which was superseded in 1797 by the National Institute of the Sciences and Arts, were

[6] *Encyclopaedia of the Social Sciences*, ed. by Edwin R. A. Seligman and Alvin Johnson (15 vols., New York, 1930-1935), XII, 19; D. Seaborne Davies, "Acontius, Champion of Toleration, and the Patent System," in *Economic History Review*, VII (1936-1937), 64; Vierendeel, *Histoire de la technique*, 6. The French theory of "the inventor's right as a natural right" has been widely adopted by other countries in their patent systems.

For the English date and the reference to Acontius, I am indebted to Professor Charles D. O'Malley of Stanford University.

[7] Tuetey, *Répertoire générale*, IV, 259; *ibid.*, VII, 259.

[8] Ballot, *Machinisme*, 25-26.

[9] *Ibid.* Ballot wrote that the recommendations for awards "were little adopted"; after examining the numerous cases described in Tuetey, *Répertoire générale*, vols. III and IV, however, I consider this an overstatement.

Berthollet the chemist, Leroy the physicist, and Vandermonde the mechanician.[10]

One of the Bureau's early actions was to request of the minister of the interior a complete chronological list of all inventions and discoveries for the preceding twenty years, as recorded by that ministry.[11] The list was to include the name of the inventor in each instance and what government recompense if any was made him. The minister of the interior forwarded this request to Tolozan, former intendant of commerce, who was in charge of liquidating the old administration of commerce.[12] Obviously the Bureau wanted to have at its disposal information concerning inventors whose periods of government protection and favor were not expired and concerning future claims for protection and favor.

With the creation of patents came the gradual establishment of a depot for records and models of inventions. The process began in 1782 when Vaucanson died, leaving to the king his excellent private collection of machines, mostly for spinning and weaving. Vaucanson had housed this collection since 1775 in the Hôtel de Mortagne, where it was open for public inspection. Its chief use, however, was in the training of workmen. Shortly after the collection came into the hands of the king, the government bought the Hôtel de Mortagne and engaged Vandermonde, a member of the Academy of Sciences, as custodian.[13] Vandermonde rapidly added models of several hundred British and Dutch machines.

This limited and highly select government depot was therefore already in existence when a second patent law, March 29, 1791, called for the establishment of a general depot of inventors' models and records, to be designated the Directory of Brevets of Invention. The enterprise was to be under the supervision of the minister of the interior.[14] No funds for the purchase of a

[10] Tuetey, *Répertoire générale*, III, xlii-xliii.
[11] *Ibid.*, IV, 258.
[12] *Ibid.*, VII, 261. This correspondence was in January, 1792, when Leroy was chairman of the committee.
[13] A portion consisting chiefly of curiosities was sent to the Academy of Sciences; the remainder at the Hôtel was more utilitarian, exemplifying textile machines in use. Frederick B. Artz, *L'enseignement technique en France pendant l'époque révolutionnaire, 1789-1815* (Paris, 1946), 25. See also Bertrand, *Académie des Sciences*, 322; Wolf, *History of Science*, 41.
[14] Art. 2. This law merely supplemented that of January 7, 1791, setting

THREE AUTOMATA OF JACQUES VAUCANSON

These automatic figures—a tambourine player, a duck, and a flute player—were great public attractions in the middle of the eighteenth century.

(Illustration from Vaucanson, *Le mécanisme du flûteur automate*, frontispiece, courtesy of the Library of Congress)

building or for its administration, however, were passed, and it remained for the National Convention in 1794 to take more definite action.

By decree of October 13, 1794, the Convention ordered the establishment at Paris of a great collection of machines and mechanical devices of all types. This assemblage was in the future to receive models of all French inventions. The Hôtel d'Aiguillon, where a great mechanical exhibition had been opened in 1793, was to house the collection, and its exhibit made the nucleus. A staff of four experts was provided to demonstrate the contents to the public. This collection rapidly increased as inventors' models came in and as gifts were made. Among the donations were the important collections of clocks and watches by Berthoud, of physical apparatus by the Nollets, and of chemical equipment by Lavoisier.[15]

In 1798 the need for larger and better housing facilities was seen. There were then three separate exhibits, one at the Academy of Sciences, a second at the Hôtel de Mortagne, and a third at the Hôtel d'Aiguillon. Accordingly the government of the Directory, by a law of June 10, 1798, set apart the buildings of the old Benedictine priory of Saint-Martin-des-Champs as the place where the three existing collections should be brought together. In the next year the transfer was carried out, and one more step was taken toward making the French collection of the Conservatory of Arts and Trades, "the largest industrial museum of Europe."[16] Wolf calls it "the first museum of Science and Technology," and says that it "may reasonably be credited with having stimulated the foundation of similar institutions elsewhere."[17]

The influence of this great institution has been immense. Not only did it house an important technological collection, but it included also an important technological library, and in the early

---

forth detailed regulations for the filing of inventors' claims to recognition. *Moniteur universel*, VII, 761-62.

[15] Only a part of Lavoisier's equipment was obtained, however. Artz, *Enseignement technique*, 25-26.

[16] The moving figure in promoting it throughout the 1790's was the Abbé Grégoire. *Ibid.*, 24-25.

[17] Wolf, *History of Science*, 42. See also Artz, *Enseignement technique*, 27.

nineteenth century it was the forum where courses of lectures by scientific experts were given. The French public and foreigners visited it in great numbers. Among nineteenth-century French inventors who attended its lectures were Jacquard and Schneider, the former the inventor of the famous silk loom bearing his name and the latter the director of the metallurgical plant of Creusot.[18] The story of its influence, however, lies in nineteenth and twentieth-century history.

The creation of a pension system and the establishment of a depot of models and records were in themselves attempts by the government to encourage invention. Other methods, too, were taken, such as awards, reimbursements, grants-in-aid, and pensions. Most of them had long been in existence. Back in 1716 the French government offered 100,000 livres for the best method of determining longitude at sea. It was but following the course of the Spanish, Dutch, and British governments, which in the period 1598-1714 had offered large prizes for the same thing.[19] Thenceforth throughout the century the French government offered a veritable stream of prizes, increasing in number as time went on, generally under the auspices of the Academy of Sciences (which to all intents and purposes was a royal scientific council), but after 1731 also through the Academy of Surgery. The sums offered, however, never again approached 100,000 livres.

Turgot as minister (1774-1776) was particularly active in offering prizes and subsidies to encourage invention. Mention has already been made of the offer of 4,000 livres for the best paper on a quicker and better method of manufacturing saltpeter. Sixty-six papers were submitted, and the award went to Thouvenel, commissioner of powders at Nancy. Two minor prizes of 1,000 livres each were also given.[20] Turgot subsidized the scientist Duhamel du Monceau in his researches on steel manufacture, and to others he made subsidies for various projects, such as the refining of sugar, the uses made of coal in Britain, improvements in textile manufactures, and a means of breaking ice floes at the

---

[18] Artz, *Enseignement technique*, 26-27.

[19] Usher, *History of Inventions*, 284; Bertrand, *Académie des Sciences*, 188.

[20] Turgot originated the contest, but it was not ended until 1782, several years after Turgot had fallen. Turgot, *Oeuvres*, IV, 376-78.

junction of the Marne and Seine. He was interested also in preventing the pollution of streams and in obtaining a better cable for ships, and he made plans to offer prizes in both matters.[21]

Although Turgot was thrown from power in 1776, his successors continued his zeal throughout the century. In 1785 the government gave several gold medals to encourage industrial development. One went to Reveillon for his contribution to the art of manufacturing wallpaper in long strips.[22] Numerous pensions were given inventors, as may be observed in the *Etat nominatif des pensions*, drawn up in 1791 and since printed in the collection *Archives parlementaires*.[23] The policy of the royal government was followed by the provinces and municipalities which also gave prizes, subsidies, and pensions to inventors.[24]

Not infrequently the government aided families of inventors in distress, especially during the Revolution. As an example, the locksmith and inventor Georget (who already had received a gold medal from the Academy of Sciences) was granted by the Legislative Assembly in 1792 the sum of 6,000 livres, partly in recognition of his achievements and partly because his large family was in need. In the same year a gift of 300 livres was made to the aged and needy daughter of a certain Porro, a deceased researcher on hemp.[25]

---

[21] *Ibid.*, 650; Douglas Dakin, *Turgot and the Ancien Régime in France* (London, 1939), 203, 252. Turgot also gave encouragements for a number of agricultural tools and implements, for a new rat trap, for a method of distilling sea water, and for means of irrigating land. *Ibid.*, 85-86, 127, 201.

[22] The medal was lost in the riot of April, 1789, when Reveillon's home was pillaged. The king had wanted to replace it, but he was aware that it could not be done at that time. In 1792, Roland, minister of the interior, proposed to the Legislative Assembly that another medal be given Reveillon. *Archives parlementaires*, XLII, 503.

[23] *Ibid.*, vols. XIII-XV.

[24] Numerous references to them can be found in *Archives des Bouches-du-Rhône, Archives de la Côte d'Or, Archives de la Gironde*, and *Archives de l'Ille-et-Vilaine*. As an example of municipal gifts, the Chamber of Commerce of Marseilles in October, 1765, voted Louis Delmas "an annual gratification of 150 livres," tantamount to a pension, in recognition of "a new plunging machine" he had invented; upon his death shortly afterward it voted the purchase of the machine from his widow, offering 450 livres for it. *Inventaire des archives historiques de la chambre de commerce de Marseille* . . . , ed. by Octave Teissier (Marseille, 1878), AA, art. 81. Many of the French cities have published inventories of their archival deposits, as have also the departments, in which one finds occasional mention of inventions.

[25] Tuetey, *Répertoire générale*, VII, 265, 268. For other cases in the same year see *ibid.*, 267, 268.

Not only did the government prior to the Revolution commonly make grants to the Academy of Sciences to be used for prizes and aid to inventors, but it also made respectable grants to certain private societies for the fostering of invention and industry. To a society of this type at Rouen it allocated 30,000 livres, and to another at Amiens it gave 180,000.[26]

The government encouraged these societies to import English textile machinery, notably the spinning machines of Arkwright and Crompton. It encouraged also, by direct and indirect means, the importation of Englishmen capable of demonstrating these machines. Its policy of adopting English textile methods and machinery dated back at least to the ministry of Fleury (1726-1743), but during the second half of the century it was accentuated steadily and several scores, possibly hundreds, of British subjects were induced to come to France, some as foremen but the larger number as common workmen, bringing English methods of manufacture in the textile and hardware industries.[27] While the greater number were English, there were also Scots, Irish, and Welsh. John Kay (1745) was the most outstanding of the earlier Englishmen to come, but the man of greatest significance appears to have been John Holker (1719-1786), a Lancashire Jacobite who escaped from prison in England after the uprising of 1745. Coming to France, he set up textile manufactures, a pottery, and a sulphuric acid plant (the first in France) at Rouen. From 1755 to 1786 he was an inspector general, having governance chiefly over foreign manufacturing establishments in the country. Probably his greatest achievement was his enticing, through certain British and French agents (one of them his son), large numbers of skilled English workers, above all from his native Lancashire. Among the more outstanding of those he brought to France were James Milne and his son Thomas and son-in-law Foxlow, who from their entry in 1779 until the early 1790's were active in establishing English machines and methods in various cities of northern France, especially Normandy. Foxlow in 1792

---

[26] These two grants were in 1788 or 1789. Ballot, *Machinisme*, 16.

[27] Rémond, *Holker*, 19, gives the number as perhaps hundreds. Rémond, who has made an exhaustive study of his subject based in no small degree on the archives, gives thumbnail sketches of several of the more prominent English immigrants. *Ibid.*, 18-21, 36-38, 41, 73, 98-99.

was in charge of a spinning establishment at the Château de la Muette (Paris). Others prominent were an Irishman named Mac-Carty, who carried to Picardy in the mid-century British methods of finishing fine woolens; Alcock, who set up at Saint-Omer, Charité-sur-Loire, and Saint-Etienne factories for turning out textile machinery, and who acted as a leading agent in bringing English workers to France; John MacCloud, a Scot, who at Holker's direction established at Amiens, Abbeville, and Sens some manufactures of fine muslins, and manufactures of piqués at various points in Champagne; Philemon Pickford, who introduced Crompton's "mule" at a number of cities in northern France; two Irish brothers named Mallois, who came to France in 1788 and erected their looms at Louvois; and Christopher Potter, an English politician and industrialist who came to France in 1789, established potteries, and acquired a reputation for large-scale manufacture on the one hand and for skill at painting glass and pottery on the other. He claimed to be the introducer of painted glass and pottery into France, but this will hardly bear examination. Virtually all of these immigrants received aid from the French government directly or indirectly. An idea of the magnificence with which some were treated can be seen in the pension of 6,000 livres given the Milnes in 1789 for their work, as also in the treatment of Pickford who in 1791 received from the government not only his lodgings, his workshop, and a grant of 6,000 livres, but also 300 livres for each of his machines. These handsome largesses bespeak the interest of the government in the country's industrial and commercial welfare.[28]

This policy of persuading foreign workmen to come to France, of course, harked back to earlier times, notably to the days of Colbert. During the 1700's the government showed eagerness not only to obtain English artisans, machines, and tools, but also to a lesser degree those of other countries. Workmen and tools from various parts of Europe were imported throughout the century. As examples may be cited a group of Saxons brought to

[28] *Ibid.*, 18, 19, 22, 99, 100, 101, 114, 115; *D.N.B.*, IX, 1026; *ibid.*, XVI, 214; Tuetey, *Répertoire générale*, III, xliii-xliv, 540-41, 546; *ibid.*, VII, 277-86. Holker was paid well indeed, receiving at first a salary of 8,000 livres, later raised in 1764 to 12,000 livres. In addition, all his traveling expenses were paid. Rémond, *Holker*, 102.

Héricourt, Burgundy, around 1740 for work in textiles; some Dutch foremen and workers brought to Amiens, Auch, and Roubaix for fabrication of velvets; and some Swiss ribbon-workers induced to settle at Nancy and Marseilles. From Greece, Constantinople, and Persia came workmen teaching Near Eastern methods of spinning thread. Even from China and India workmen and spinning wheels were imported in the 1740's and 1750's. In 1752 Trudaine de Montigny brought in a group of eighteen East Indian weavers with their looms to demonstrate their techniques to French workmen. Due to sickness and homesickness, this colony did not remain long in France; moreover, the East Indians were perhaps more of a nuisance than an asset inasmuch as they wrought much destruction to the château where they worked; the government in time (1788) felt obliged to reimburse the owner 2,500 livres for the damage.[29] At length in the 1780's a second group of fifty-two East Indians were imported from Malta, where they had migrated, and were settled at Thianges, near Meaux, where for a period they engaged in weaving, then returned to the East.[30] Thus the government was reaching in various directions in an endeavor to obtain better equipment and techniques.

The government attempted to teach the people not only to use English machines but also French machines and methods. In the last decades of the century it sent out agents to instruct the people of the provinces on the best ways of handling silk.[31] On the eve of the Revolution it placed one of Brizout de Barneville's machines for spinning "superfine muslins" like those of India at the institution of the Quinze-Vingts for the purpose of instructing students.[32] Machines by Lhomond, Pickford, and the Milnes likewise were placed there for the same purpose.

In short, the government was active in various ways in its endeavor to encourage invention and new ways of manufacture, with the goal of improving the economic welfare of the country. Its efforts were supplemented by those of individuals and societies

---

[29] Rémond, *Holker*, 16-19; P. Boissonnade, "Trois mémoires relatifs à l'amélioration des manufactures de France sous l'administration des Trudaines (1754)," in *Revue d'histoire économique et sociale*, VII (1914-1919), 60. Eleven of the Indians worked with silks and seven with cottons. Five were slaves.

[30] Rémond, *Holker*, 35.

[31] McCloy, *Government Assistance*, 270.

[32] Tuetey, *Répertoire générale*, III, xliii.

of encouragement. Occasionally gifts as prizes were offered by them, as in the instance of a large legacy in 1714 by Rouillé de Meslay, a "councillor to parlement" in Paris, providing for two prizes to be given each year by the Academy of Sciences, one of 2,000 livres for a treatise on the movements of the planetary system, and a second of 1,000 livres for the shortest and best method of reckoning longitude at sea. The legacy carried 100,000 livres as principal, and its annual income was reckoned at 4,000 livres. The son unsuccessfully contested the will.[33] Similarly in 1789 a gift of 12,000 livres was made to the Academy of Sciences by the banker Germain, to be offered as a prize for a better hydraulic machine to replace those at the Pont Neuf and the Pont Nôtre-Dame in Paris.[34]

Beginning in 1776 several societies for the encouragement of invention and industry were founded. The first, with the long title "The Free Society of Emulation for the Encouragement of Inventions Which Tend to Perfect the Application of the Arts and Trades in Imitation of That of London," was organized in 1776 at Paris by the Abbé Beaudeau. In order to create prestige for such societies at that time, it was necessary to have the sponsorship of an imposing array of the socially prominent. The Free Society was not different. To its membership belonged twenty great lords, four noble ladies, nineteen lesser nobles (judges in the courts), in addition to a number of bankers, merchants, architects, savants, and others of the bourgeoisie. When it was dissolved in 1781, it had disbursed 2,300 livres in prizes and 14,964 livres in other forms of encouragement.[35] Similar bodies grew up elsewhere: one at Rheims in 1778, another at Rouen in 1787, and a third at Amiens in 1789. The last two were organized with government encouragement and received government funds, as already mentioned.[36] There were, in addition, in the early 1780's three "museums," or societies of education and encouragement, in Paris, which offered lecture and laboratory courses under the direction of savants to students interested in what today would

[33] Bertrand, *Académie des Sciences*, 176-83.
[34] Tuetey, *Répertoire générale*, III, 98. In the contest of 1789, however, no prize was awarded, as the Academy did not consider any of "the works presented" to have sufficient merit.
[35] Ballot, *Machinisme*, 15.
[36] *Ibid.*, 16.

be called technological training. The most important of the three was directed by Pilâtre de Rozier, later to acquire renown in ballooning. Among those lecturing in 1786 were Garat, La Harpe, Condorcet, Fourcroy, and Monge—men of high merit.[37]

In addition to gifts, prizes, and instruction, these societies used journalistic advertisement and were active in the importation of English machinery. Two journals arose in Paris during the Revolution for the promotion of invention: the first, originated in 1790, was entitled *Annales instructives, ou journal des découvertes en tout genre*, and the second, begun in the year III (September, 1794–September, 1795), was termed *Journal des inventions et découvertes*, owned and directed by the Lycée of the Arts, Pilâtre de Rozier's old "Museum," whose name and scope of activity had met some changes since his death.[38] The latter journal was not financially self-supporting, due to an insufficient number of subscribers, and the government twice made it appreciable subsidies.[39]

Certain of these societies were active in introducing English technicians and machines into France. The society at Amiens sent two Englishmen to England in search of a "mule-jenny" by Crompton, while that at Rouen succeeded in early 1788 in obtaining for the city "twelve large jennies, twenty-four small ones, a carding machine, a hundred spinning wheels for flax"; and for the rural communities, "twelve large jennies, five hundred small ones, six carding machines, and two thousand spinning wheels." It hired the English technician Garnett to construct carding machines, paying him large sums, and it attempted to get English machines through other agents.[40]

All in all, France had a total of nine hundred spinning jennies in 1790.[41] Crompton's "mule" and Watt's steam engine were just beginning to make their appearance in France.[42] It should be borne in mind that the English had an embargo on the export of these machines, and it was exceedingly difficult to obtain

[37] *Ibid.*, 15. Still other societies existing in the 1790's are named in Schmidt, "Industrie cotonnière," 46.
[38] *Moniteur universel*, IV, 320; Schmidt, "Industrie cotonnière," 47 n. 1. Lecturing in the Lycée at this period were Lavoisier, Berthollet, Darcet, Daubenton, Lalande, and Fourcroy.
[39] Ballot, *Machinisme*, 29.
[40] *Ibid.*, 16-17.
[41] Schmidt, "Industrie cotonnière," 26-27.
[42] *Ibid.*, 27; Ballot, *Machinisme*, 17. Ballot credits the Périers with first introducing Watt's engine in France.

them.[43] Those caught in the act of transporting the newly invented machines were subject to imprisonment and fine, and the safest step was to have a skilled English mechanic reproduce them on French soil. The sums offered had to be large, for the Englishman ran the risk of not being able to return to his country without facing the penalties of the law. One Harding and an assistant who in 1798 or 1799 were arrested for transporting English textile machines and workers to France were condemned to twelve months in prison and a fine of £500.[44]

What was the attitude of the French worker toward this enthusiasm for invention and English machinery? Here the scene turned macabre. During periods of prosperity the worker perhaps shrugged his shoulders in indifference. But it became otherwise in the period of the great famine and economic depression which gave birth to the Revolution. From 1788 to 1791 riots occurred in the industrial cities where English machinery had been installed, and most of the machines were destroyed by mobs who reasoned that they were in part responsible for their unemployment. These "Luddite riots" began at the industrial city of Falaise on November 11, 1788, where all the new spinning machines that could be found were broken. They spread to Rouen, one of the leading centers of French industry, where in July, August, and October, 1789, enormous damage was done. All types of machinery were wrecked—spinning machines, carding machines, looms, and so on. More than three hundred jennies were destroyed, six carding machines, and a number of machines of Barneville, representing a loss exceeding 100,000 livres. The rabble in its blind rage had destroyed in a few days of vandalism all that its society of encouragement had done in a constructive way.[45]

The riots spread to other places. At Saint-Etienne on September 1, 1789, workers attacked machines which made pitchforks (an invention of Sauvade) and completely destroyed them.

---

[43] Between 1773 and 1786 Parliament passed five laws "to prevent the export of the new tools and machines, or of plans and models, and had forbidden the emigration of skilled artisans." H. Heaton, "Industry and Trade," in A. S. Turberville (ed.), *Johnson's England: An Account of the Life & Manners of His Age* (2 vols., Oxford, 1933), I, 243.

[44] Schmidt, "Industrie cotonnière," 45.

[45] Ballot, *Machinisme*, 20-21; Schmidt, "Industrie cotonnière," 53. Ballot gives the number of jennies destroyed at Rouen as more than seven hundred.

At Lille in 1790 workmen forced the removal of spinning and carding machines. At Paris in the same year complaint forced the closing of certain charity spinning workshops employing English machines.[46] In 1791 agitation developed against the use of jennies at the Quinze-Vingts, and in the face of threats to burn them the government had to remove them to the home of inspector general Brown. At Roanne spinning machines were destroyed. Even in 1792, when economic conditions had become better, ribbon workers of Paris went before the Legislative Assembly with the demand that all mechanical looms in the country be destroyed, claiming that they were responsible for the unemployment of 100,000 persons.[47] Nor do these appear to be the only places where fury was displayed. At Troyes, for example, several groups of workers rose in indignation against the machines that destroyed employment.

Little did the rioters realize what they were doing. They were setting back the clock of time, at least industrially, some two decades; for it was not until after the wars of the Revolution and Napoleon that France was able to introduce such machines again. Meanwhile England had the advantage of this machinery in the long conflict. In a struggle where manpower was so short and skill counted for so much, France was to suffer terribly for this blind folly.

In conclusion, it may be said that the efforts of the government, of individuals, and of private societies to promote invention and industry bore impressive results. There was keen competition for the medals and prizes offered, and on occasion the invention or discovery was a notable one. No doubt these awards had something to do with channelizing the efforts of many inventors, and much to do with the creation of public opinion favorable to invention. Even more significant than the prizes and medals, however, were the subsidies and pensions. After all, only a minor number of the inventions of the century were inspired by prizes. A much larger number came as the result of subsidy or pension,

---

[46] It was estimated that one person using an English jenny could turn out as much work as six hand workers, and produce thread of a better quality. Tuetey, *Assistance publique*, II, 510-11; Ballot, *Machinisme*, 21; Schmidt, "Industrie cotonnière," 53.

[47] Schmidt, "Industrie cotonnière," 53; Ballot, *Machinisme*, 21-22, 271.

with the inventor working on a project of his own choice. Investigation reveals that the most prominent inventors were the authors of several inventions and the recipients of pensions which freed them for inventive activity. This appears to have been the most desirable form of encouragement, and after it, the grant of subsidies and reimbursements, which likewise served economically to enable the inventor to apply himself to his inventive work. The vast majority of the inventors, it may be added, were of bourgeois origin, and few were in economic circumstances sufficiently independent that they could long work at invention without bringing hardship upon themselves; and yet many died in penury. It is to the credit of the French government and people of the eighteenth century that they did assist many inventors; it is regrettable that they did not assist many others.

Conclusion

MANY QUESTIONS NO DOUBT HAVE BEEN PROVOKED BY READing of the foregoing chapters. What was the education of the inventors? What motives were uppermost with them? What section of France furnished the most inventors? How did France compare in invention with other countries? Unfortunately the information concerning most inventors is meager; of some little more than their names is known. Even the most famous have received treatment only too scanty, for biographers have looked elsewhere for their subjects. It is therefore upon a limited number of the most prominent that observation must be based for answers.

None of these questions carries more interest or importance than that of the inventors' education. Possibly no class of citizens contributes more toward the welfare of a nation than its inventors. A government therefore cannot afford to neglect training its youth along channels most propitious to creative thought and activity. Careful analysis shows that most of the French inventors of the eighteenth century were men with good minds, variously trained. Relatively few except the physicians had received a university education. Others had attended colleges maintained by the religious orders. A small number had been educated at technical schools, more especially in engineering. Then there were those trained by apprenticeship, in which group fell nearly all the surgical and textile inventors and many in other fields. This number was by far the largest of the four.[1] The elements of inquiry

---

[1] The type of education that prominent inventors received is listed below.
University: Etienne Montgolfier, Berthollet, Réaumur, Macquer, Bordeu, Bichat, and Pinel.
College: Joseph Montgolfier, Vaucanson, Camus, the Abbé Chappe, and the Abbé Mical.
Technical school: Carnot, Cugnot, D'Arçon, and apparently Lebon and Berthelot.
Apprenticeship: Coulomb, Conté, Mlle. Biheron, Mme. du Coudray, Baseilhac, Daviel, Desault, J. L. Petit, Fauchard, Larrey, Lasalle, Barneville, Jouffroy, Carcel, the Leroys, De Cour-

CONCLUSION

and research appear to have been imparted more by this mode of training than by classroom lecture and quiz.

In observing that the largest number of inventors were men and women trained through apprenticeship, it should be recalled that in the eighteenth century more French youth were trained for lifework by apprenticeship than by universities, colleges, and technical schools, although the number attending these last was surprisingly large.[2] It seems clear, therefore, that all forms of education in the century contributed their share toward the training of inventors. Few if any inventors had received training in the humanities (or classics) alone; they had received instruction also in certain of the sciences. And this, it appears, was vital. Invention bloomed out of a close association with and careful training in some field of work that was the inventor's specialty or hobby.

The age at which inventors made their contributions could be plotted on a curve with the age of forty at the zenith. Nevertheless, some of the most interesting inventions were made by men in their twenties and in their fifties and sixties.[3] Many inventors produced but a single invention, or at least only one of consequence. On the other hand, the list of those contributing more than one invention was surprisingly large.[4]

As for native locale, Paris was the birthplace of the largest number of inventors, perhaps a third of the whole; it was followed more closely by Rouen than any other city. Bordeaux and Gren-

---

tivron, the Marshal de Saxe, Gribeauval, and most of the inventors with a military background.

[2] McCloy, *Government Assistance*, 413, 417-22.

[3] The age group of men when they made their initial inventions is listed below.

Twenties: Vaucanson, the Abbé Chappe, Bichat, and De Courtivron.

Thirties: Etienne Montgolfier, Charles, Jouffroy, Argand, Robert, Berthollet, Macquer, Lebon, Camus, Beaumarchais, Pereire, and Bordeu.

Forties: Joseph Montgolfier, Réaumur, Cugnot, Coulomb, Conté, Berthe-lot, D'Arçon, and Pinel.

Fifties and sixties: Lasalle, Carcel, and Daviel. If subsequent inventions were included, the number of those making contributions in the fifties and sixties would be increased.

[4] Jouffroy, Chappe, Robert, Barneville, and Larrey are among those making a single invention. Those contributing twice include Joseph Montgolfier, Charles, Argand, Vaucanson, Conté, Camus, Périer, Coulomb, Berthollet, Berthelot, Macquer, Lasalle, Lhomond, Cugnot, Mical, Pereire, Fauchard, and J. L. Petit. Some of these men produced several inventions.

oble appear to have made but a single contribution each. It would seem that Lyons would have ranked high in invention, inasmuch as it was the center of the silk industry, and early in the next century it did make a stellar contribution in Jacquard's loom, but there exists apparently no record of a single notable eighteenth-century inventor born there. The greater number of French inventors in the 1700's were born not in the large cities but in the towns, villages, and rural districts . Most of them later went to the larger cities, especially to Paris, to complete their education and to live. Not a few country boys from the far-off Alps and Pyrenees came to Paris to play a spectacular role in furthering the cause of science and humanity for their country. As for the regions producing these men, Paris and its neighboring territory stood highest by far; next came the Rhône valley and the Alpine section of southeast France; third was the center and southwest; a close fourth was the western region of Normandy, Brittany, and Flanders; and trailing all came Alsace-Lorraine and the adjacent territory of northeastern France, the natal homeland of Cugnot and Camus. The barrenness of this last region is surprising in view of the fact that even then Alsace-Lorraine was an industrial, mining region; but in partial extenuation it can be pointed out that Lorraine did not become an integral part of France until 1766 on the death of its king Stanislas. Roughly speaking, it can be said that all parts of France produced the men who were to become inventors in the 1700's in proportion to their population. Moreover, every class from peasant to noble contributed, with the middle class furnishing much the largest number and the nobility the least. As filings to a magnet, so most inventors were attracted to Paris, which became the home of fully two thirds of them.[5]

[5] The birthplaces of the leading inventors are listed below.
Paris: Robert, Macquer, Beaumarchais, Carcel, Périer, Mlle. Biheron, Pierre Leroy, and J. L. Petit.
Rouen: the elder Barneville and Le Cat.
Bordeaux: Pereire.
Grenoble: Vaucanson.
The Alps: the Montgolfiers, Bichat, Lasalle, and Berthollet.
The Pyrenees: Bordeu and Larrey.
Rural areas: Charles, Cugnot, Carnot, Berthelot, Pinel, Desault, Conté, Camus, Chappe, Lebon, Coulomb, Réaumur, and Fauchard.
A crescent drawn on the map of France through Paris, Grenoble, and Toulouse, with one tip at Rouen and the other in the Pyrenees, would represent the most fertile natal field for inventors.

THE EXTERIOR OF A SEMAPHORE TELEGRAPH TOWER
(Illustration from Figuier, *Merveilles*, II, 53, courtesy of the Library of Congress)

CONCLUSION

The question of the motives or incentives of the inventors is much more difficult to solve. In our own century, due in no small degree to the writings of Karl Marx, a reader might easily jump to the conclusion that the economic motive was paramount. With not a few of the inventors it must have been. It is probable, in fact, that most of them hoped to realize some financial benefit from their inventions. This is far from saying that the hope of economic gain was the paramount motive, or indeed that it was the original driving force. Few inventors benefited appreciably from their inventions; a much greater number squandered their inheritance and savings on their inventive activity. The largest return to most of them was a government pension, usually modest. Some received no reward whatever. As a matter of fact, some inventors were so indifferent to monetary returns that they renounced claim to economic exploitation of their inventions. In this category were Berthollet, Berthelot, Camus, and Saint-Sauveur. With difficulty the friends of Conté persuaded him not to do likewise, and only the consideration of the other members of his family moved him. Vaucanson bequeathed his collection of machines, on which he had spent much of his earnings, to the king for public display. With these men patriotism and humanitarianism burned brightly. Even more brightly burned the desire for achievement and fame; this was the dominant motive of the French inventors. Economic returns were of secondary consideration, and humanitarianism and patriotism were seldom absent.

Some few inventors, no doubt, approached their work incited by prizes offered for some particular invention, but the greatest number selected their own field or problem. They appear to have caught a gleam of their invention, perhaps at some casual moment, and to have labored at it for months or years. In most cases they observed a mechanism doing its work slowly or awkwardly and quickly saw how it could be bettered. The physicist Charles heard of the balloon at Annonay raised by hot air; at once he thought of the superiority of the newly discovered hydrogen for this purpose. Vaucanson has related that the idea of his flute-player came to him when as a young visitor to Paris he stood watching the marble statue of a flautist in the Garden of the Tuileries. Years passed before he put his vision into effect, after long labor and much discouragement into which the attempts of relatives to dissuade entered. Possibly this did not happen in all cases or even in the majority of cases, and yet it seems plausible that there was usually first a vision, then a hope, and then months of unremitting toil.

The government recognized the importance of invention and made an earnest endeavor to encourage it. It probably did not do this sufficiently, however. No inventor was paid as much as a provincial intendant, an army general, an archbishop, or a member of the royal family; nevertheless, his service to France was often greater. Frequently, of course, it was not possible at the time to realize the full value of the inventions, however flattering or critical the report of the investigating committee of the Academy of Sciences or later of the Bureau of Consultation of the Arts and Trades. Normally, inventors today are given only patents and the rights to exploit their inventions for a fixed number of years. In eighteenth-century France the inventor generally was allowed monopolistic rights of exploitation for fifteen years and was often granted in addition a pension or gratification, which might vary from a few hundred livres to the handsome sum of 15,000 livres given Réaumur.

There were cases of unpardonable neglect, as those of Jouffroy and Cugnot, due to jealousy, lack of insight, change of government personnel, and so on. In general, however, the French government appears to have followed a rather creditable policy toward inventors and inventions. The criticism that more inventions did not result must be charged largely elsewhere. Much of the invention of the nineteenth and twentieth centuries has been encouraged and financed by large industrial establishments, research foundations, or wealthy universities. Little of this support was available in eighteenth-century France. Generally the inventor expended his own savings and patrimony. Cases of large industrial establishments like that of Sèvres which gave leisure to Macquer or that of Jouy which provided Widmer with leisure and funds were rare. Not until the Germans in the late nineteenth century caught the vision were industrialists to realize the importance of having staffs of chemists and physicists for the creation of new devices and processes. Private business of eighteenth-century France thus was more to blame than the government for failure to pursue invention. The business leaders, moreover, were slow to organize societies to promote invention. Such private societies did not develop until the 1770's. In England, which excelled in certain fields, notably textiles, hardware, and agricultural machinery, a private organization entitled the Society for the Encouragement of Arts, Manufactures, and Commerce had been founded in 1754. It distributed large sums in prizes and medals to inventors. The English public became aware earlier than the French by some three decades of the importance of invention and the need for its support. As for the universities, neither the British nor the French institutions in the 1700's appear to have had the funds or the interest to devote their time to invention. They were dominated by classical studies, and research received little attention. Yet in that age of Latin-speaking professors and students a remarkable university came into being in the 1730's at Göttingen, Germany, devoted to research and the seminar method, and destined to work a great revolution in educational trends. If invention has made greater strides in the nineteenth and twentieth centuries, it has been largely because education has been more popular and more scientific, because universities and technical schools have had greater financial resources at their command, because great research foundations have arisen, because industry has become convinced of the necessity of maintaining corps of scientific research experts, and because the general public is more enthusiastic for invention.

This is not to say that capital for industrial purposes was altogether lacking during the century. On the contrary, it is surprising to what degree wealth was invested in French industry and mining, and how many were the large-scale establishments. Some were amazingly big. There were sugar factories at Cette on the Mediterranean, for instance, which necessitated the initial expense of 450,000 livres, and there were the Anzin coal mines in north France which required the outlay of 20,000,000 livres (12,000,000 for excavations and 8,000,000 for machinery) and twenty-two years of work before they were actually opened and any returns received. In these mines more than thirty shafts were sunk, some 900 feet and others 1,200 feet, and a vein of water was crossed.[6] The Oberkampf factory at Jouy in 1789 is said to have had a capital investment of almost 9,000,000 livres. A building for textile manufacture at Montauban cost 125,000 livres, and another at Louviers 200,000.[7] The metallurgical plant at Le Creusot, the glass factory at Saint-Gobain, the wallpaper factory of Reveillon at Paris, the silken goods factory of Van Robais at Abbeville, and the textile plant of Ducrétot at Louviers were only some of the more celebrated of many large establishments having heavy investments of capital and often employing hundreds of workmen. Some of these factories brought lucrative returns and their owners became wealthy. Unfortunately, we know less of those that failed.

The number of inventions in the 1700's was certainly much greater than that of the 1600's, although it would be difficult to give the proportion. Of the seven volumes containing the elaborate account of the inventions approved by the Academy of Sciences from its establishment in the 1660's to the arbitrary date of 1754, only the first treats the inventions down to 1701; the remainder deal wholly with the 1700's. It would appear, accordingly, that early eighteenth-century inventions were more than three times as numerous as those of the preceding four decades. The inventions recorded in the last six volumes of this work ran irregularly. In some years many were approved; in others, few or none. Thus there were thirteen in 1702, seven in 1705, three in 1709, one in 1712, fifteen in 1716, three in 1721, and twenty-two in 1726. During the decade 1702-1711, the academy approved 61 inventions; during the next ten years, 69; from 1722 to 1731, 99; during the years 1732-1741, 63; and in the next decade, 36. As might be expected, there was a considerable decrease in the number of inventions during the great wars of the Spanish Succession (1702-1713) and the Austrian Succession (1740-1748), when French armies were fighting on several continents. Although 67 inventions were approved by the academy during the

[6] G. Renard and G. Weulersse, *Life and Work in Modern Europe (Fifteenth to Eighteenth Centuries)* (New York, 1926), 189.

[7] Henri Sée, *Economic and Social Conditions in France during the Eighteenth Century*, tr. by Edwin H. Zeydel (New York, 1927), 166-67.

earlier war, 26 of them came in 1702-1703 before the burden of the struggle made itself felt; in subsequent years the number fell, as two in 1708, three in 1709, four in 1710, three in 1711, one in 1712, and four in 1713. The number of inventions was more uniform during the second war despite its severity, although in 1740, 1743, and 1747 only two were recognized each year. Only three were approved in 1748, and none in 1749 or 1750. Between these two wars France was engaged in a third conflict, though of smaller proportions, the War of the Polish Succession (1733-1738). One again the number of inventions fell. Where 27 were approved in 1733, only four were approved in 1734, three in 1735, two in 1736, two in 1737, and none in 1739-1740.

Unfortunately editor Gallon's work was not carried beyond 1754. The *Mémoires* of the Academy of Sciences, however, continued to carry annual lists of machines and inventions approved by that body to about 1773, and when the Institute of France was organized in 1794, it resumed the practice. Some doubt may be raised about the accuracy of the listings, since variance exists between Gallon's figures and those of the *Mémoires* prior to 1754; but the discrepancy is not great. The figures certainly are the most accurate available for the period, and from them it can be seen that the number of inventions ran slightly higher for the second half of the century than for the first half. Thus, the *Mémoires* list 69 inventions for the 1760's, 19 for the two years 1770-1771, and 24 for the three years 1796-1798.

It should be borne in mind that early in the Revolution custody of patents was put directly in the hands of the Bureau of Consultation of the Arts and Trades rather than the Academy, but there was a close relationship between the membership and working of the two bodies. The work of the Bureau for 1789 to 1792 is well summarized in Volumes III and VII of Tuetey's *Répertoire générale des sources manuscrites de l'histoire de Paris pendant la Révolution française*, and it would appear from this listing that the 1790's brought an increase in the number of inventions.[8]

[8] Tuetey, *Répertoire générale*, lists eight inventions for 1789, four for 1790, seven for 1791, and 69 for 1792, making a total of 88 for these years. The editor, moreover, deals only with Paris and apparently does not take cognizance of inventions made in the provinces.

CONCLUSION

Not only did the second half of the century see a larger number of inventions than the first half, but in this latter period there were many more inventions of real significance. This increase in number and significance continued in the nineteenth and twentieth centuries, when French inventions have been many and their value often enormous. Niepce, Daguerre, Pasteur, and Madame Curie are but a few of the names of the later inventors. These inventions continued to spring from the demands of the times and the developments in pure science.[9]

The tie was close between eighteenth-century interest in invention and the Enlightenment and the humanitarian movement of the time. Several of the French *philosophes* of the period wrote enthusiastically of invention and scientific advance as steps in the betterment of man's living conditions. Imbued with the idea of progress in the modern world, as the late Professor Bury has shown,[10] they ranged themselves on the side of the Moderns as zealots of science and social improvement in the Controversy of the Ancients and the Moderns. In this cause Voltaire was the most vocal, giving eulogistic treatment to science and invention in his *Letters on the English,* his *Age of Louis XIV,* his *Essay on the Manners and the Spirit of the Nations,* his *Summary of the Century of Louis XV,* and his poems, discourses, and private correspondence.[11] Once or more he complained that invention had been used for man's destruction rather than for his betterment.[12] He criticized France as backward in invention because of her reluctance to accept inoculation for smallpox. He was blind and splenetic enough to ridicule Réaumur as a quack after Réaumur in 1738 did not support a paper which Voltaire submitted for a prize offered by the Academy of Sciences.[13]

---

[9] A recent book dealing with nineteenth and twentieth-century inventions is Léon Guillet, *La France, pays de grandes découvertes* (Paris, 1947).

[10] J. B. Bury, *The Idea of Progress: An Inquiry into Its Origin and Growth* (London, 1920).

[11] François M. A. de Voltaire, *Oeuvres complètes de Voltaire,* ed. by Louis Moland (new ed., 52 vols., Paris, 1877-1885), XXII, 118-20; *ibid.,* XIV, 534-36, 557-58; *ibid.,* XII, 54; *ibid.,* XV, 431; *ibid.,* VIII, 439-42; *ibid.,* IV, 494.

[12] *Ibid.,* X, 188-89.

[13] *Ibid.,* XXXVIII, 145; *ibid.,* XXXIX, 321; *ibid.,* XXXIV, 476-77, 539; *ibid.,* XLII, 330. In 1737 Voltaire referred to Réaumur in complimentary fashion. *Ibid.,* IX, 401. The key to the change is found in a letter to Pitot, dated May 18, 1738. *Ibid.,* XXXIV, 467-77.

CONCLUSION

Diderot made few allusions in his writings to inventions, but he referred in complimentary fashion to a number of inventors, such as his neighbor Mademoiselle Biheron, Réaumur, Vaucanson, Papin, Julian Leroy, and Daviel.[14] His enthusiasm for science and invention was revealed chiefly in his editing of the great *Encyclopédie, ou dictionnaire raisonné des sciences, des arts et des métiers* (Paris, 1751-1772), with D'Alembert as joint editor of some of the earlier volumes. It was a great undertaking for any period, and Diderot's intention, as his daughter and biographer has related, was to create a work in writing and in plates which would leave a record for future generations as well as to inform all curious persons of his own age of the arts and crafts and culture of the eighteenth century. So well was the work received that it went through several editions and inspired Charles Panckoucke to project his colossal *Encyclopédie méthodique*, published at Paris between 1782 and 1832.

Others among the *philosophes* who pleaded the cause of science and invention were D'Alembert and Condorcet, both of whom wrote biographical sketches of members of the Academy of Sciences (published collectively in six volumes), and La Condamine and the Abbé Morellet, who wrote in favor of inoculation for smallpox.[15] Condorcet also displayed his enthusiasm for science and invention in his famous posthumous *Esquisse d'un tableau de l'esprit humain* (Paris, 1795).

The *philosophes*, however, were far from solitary champions of science and invention. Voltaire and Diderot referred to the general popularity of inventors, and there is every reason to believe that the same enthusiasm for science and invention was found among all ranks of the cultured in France, including churchmen, anti-*philosophes*, and the crown. The writings on science and invention by the *philosophes* were negligible in comparison to the output by other writers, as may be seen by turning the pages of Quérard or any other guide to French literary productivity of the century. The support of invention by capitalists

---

[14] Diderot, *Oeuvres complètes*, VI, 33; *ibid.*, I, 279, 333-34; *ibid.*, II, 18, 68, 148, 188, 283; *ibid.*, IX, 240, 373; *ibid.*, XIX, 175; *ibid.*, XX, 61-62.

[15] On all these, see J. M. Quérard, *La France littéraire, ou dictionnaire bibliographique des savants, historiens et gens de lettres de la France* . . . (12 vols., Paris, 1827-1864).

and royal officials was considerable, and reveals the popularity of science during the period.

The interest in invention was a phase of the Enlightenment and the humanitarian movement, in which not only the *philosophes* but most cultured people shared in a varying degree. Virtually all persons in France were in favor of human betterment, as can be seen from the *cahiers* on the eve of the Revolution and the journals and pamphlet literature of the time. It can also be seen in the vast correspondence during the century dealing with government assistance to the needy. Most groups believed that science and invention could aid greatly in the process, but wide differences prevailed over the steps that should be taken to bring these things about and over how fast the changes should be made.

It is customary to think of French dominance of continental Europe in the period 1794-1813 as coming from an upsurge of nationalistic feeling. So it was in part, but armies have to possess more than that to win battles. A nation has to have superiority in leadership, in strategy, in equipment. The French appear to have had all these, due in part to their inventive achievements. French gunnery and strategy, evolved by a succession of experts, displayed their superiority, aided to be sure by that remarkable new development of the Revolution, the *levée en masse,* by which all the citizenry, male and female, were conscripted either for the army or for the factories and fields. French superiority in the military did pay off, and French culture dominated Europe even more than it had in the long period since 1648. Some today would say that this military dominance was fatuous. So far as France was concerned, perhaps it was, for she dissipated her manpower and was left with little more than a memory of her day of grandeur. Yet for Europe it was otherwise, for it sowed the seeds of liberalism and nationalism that have figured so largely in Continental history since then, toppling crowns and carving new state boundaries.

In other and more direct ways the world is indebted to French inventive genius of the eighteenth century. Through the metric system, modern chemical nomenclature, the service of surgeons on the battlefield, the operations for cataract, mastoiditis, and other ailments, the world is heir to the legacy of eighteenth-century French invention. And not in these respects alone. French inventors were numerous, and many of their contributions remarkable. They help to illustrate why France at that period was great and her civilization dominant on the Continent. The French can well be proud, as indeed they are, of these extraordinary men of their past.

Bibliography

THIS BIBLIOGRAPHY INCLUDES ONLY THE PRINCIPAL WORKS used in writing the text. Bibliographical information for every work used has been given in the first citation.

Archives parlementaires de 1787 à 1789. Recueil complet des débats législatifs & politiques des chambres françaises. Ed. by J. Madival and E. Laurent. 1st ser., 2d ed., 81 vols. Paris, 1879-1913.
Artz, Frederick B. L'enseignement technique en France pendant l'époque révolutionnaire, 1789-1815. Paris, 1946.
Baas, Johann Hermann. Outlines of the History of Medicine and the Medical Profession. Tr. by H. E. Handerson. New York, 1910.
Ballot, Charles. L'introduction du machinisme dans l'industrie française. Ed. by Claude Gével. Paris and Lille, 1923.
―――――. "Philippe de Girard et l'invention de la filature mécanique du lin," in Revue d'histoire économique et sociale, VII (1914-1919), 135-95.
Bast, Amédée de. Merveilles du génie de l'homme: découvertes, inventions. Récits historiques, amusants et instructifs . . . Paris, 1852.
Baud, Paul. L'industrie chimique en France: étude historique et géographique. Paris, 1932.
Bertrand, Joseph. L'Académie des Sciences et les académiciens de 1666 à 1793. Paris, 1869.
Biographie universelle, ancienne et moderne; ou, histoire, par ordre alphabétique de la vie publique et privée de tous les hommes qui se sont fait remarquer par leurs écrits, leurs actions, leurs talents, leurs vertus ou leurs crimes. Ed. by L. G. Michaud and J. F. Michaud. 85 vols. Paris, 1811-1862.
Bondois, Paul M. "L'industrie et le commerce sous l'ancien régime," in Revue d'histoire économique et sociale, XXI (1933), 140-89.
Boutaric, Augustin. Les grandes inventions françaises. Paris, 1932.
Buck, Albert H. The Dawn of Modern Medicine: An Account of the Revival of the Science and the Art of Medicine Which Took Place in Western Europe during the Latter Half of the

Eighteenth Century and the First Part of the Nineteenth. New Haven, Conn., 1920.

Cabanès, Augustin. *Chirurgiens et blessés à travers l'histoire, des origines à la Croix-Rouge.* Paris, 1918.

Castiglioni, Arturo. *A History of Medicine.* Tr. by E. B. Krumbhaar. New York, 1941.

Caullery, Maurice. *French Science and Its Principal Discoveries Since the Seventeenth Century.* New York, 1934.

Cavallo, Tiberius. *The History and Practice of Aerostation.* London, 1785.

Chaptal de Chanteloup, J. A. *Chemistry Applied to Arts and Manufactures.* 4 vols. London, 1807.

Clouzot, Henri. *Histoire de la manufacture de Jouy et de la toile imprimée en France.* Paris and Brussels, 1928.

Coleby, L. J. M. *The Chemical Studies of P. J. Macquer.* London, 1938.

Coles, Leonard A. *The Book of Chemical Discovery.* London, 1933.

Collection de documents pour servir à l'histoire des hôpitaux de Paris. Ed. by Michel Möring and Charles Quentin. 4 vols. Paris, 1881-1887.

Condorcet, M. J. A. N. Caritat, Marquis de. *Oeuvres complètes.* 21 vols. Brunswick and Paris, 1804.

———. *Oeuvres de Condorcet.* Ed. by A. Condorcet O'Connor and M. F. Arago. 12 vols. Paris, 1847-1849.

Coulon de Thévenot, J. F. *L'art d'écrire aussi vite qu'on parle; ou, la tachygraphie française, dégagée de toute équivoque ...* New ed. Paris, 179–(?).

Davis, Charles Thomas. *The Manufacture of Paper: Being a Description of the Various Processes for the Fabrication, Coloring, and Finishing of Every Kind of Paper ...* Philadelphia and London, 1886.

Davis, Nathan Smith. *History of Medicine, with Code of Medical Ethics.* Chicago, 1903.

Delambre, Jean Baptiste Joseph. *Rapport historique sur les progrès des sciences mathématiques depuis 1789, et sur leur état actuel.* Paris, 1810.

Devaux, Paul. "L'incubation artificielle," in *Nouvelle revue,* LXXVII (1892), 564-82.

Diderot, Denis. *Oeuvres complètes de Diderot.* Ed. by J. Assézat and M. Tourneux. 20 vols. Paris, 1875-1877.

Doolittle, William H. *Inventions in the Century.* Philadelphia and London, 1903.

Ducros, Louis. *French Society in the Eighteenth Century.* Tr. by W. de Geijer, London and New York, 1927.

Dutil, Léon. *L'état économique de Languedoc à la fin de l'ancien régime (1750-1789)*. Paris, 1911.
Encyclopédie méthodique, ou par ordre de matières. Ed. by Charles Panckoucke. 229 vols. Paris, 1782-1832.
Figuier, Louis. *Les merveilles de la science, ou description populaire des inventions modernes*. 4 vols. Paris, 1867-1870.
Fontenelle, Bernard Le Bovier de. *Oeuvres de Monsieur de Fontenelle*. New ed., 10 vols. Paris, 1758.
Fournier, Edouard. *Le vieux-neuf: histoire ancienne des inventions et découvertes modernes*. 2 vols. Paris, 1859.
Franklin, Alfred. *La vie privée d'autrefois: arts et métiers, modes, moeurs, usages des Parisiens de XIIe au XVIIIe siècle d'après des documents originaux ou inédits*. 1st ser., 23 vols. Paris, 1887-1901.
Franklin, Benjamin. *The Works of Benjamin Franklin*. Ed. by John Bigelow. 12 vols. New York and London, 1904.
Garrison, Fielding H. *An Introduction to the History of Medicine*. 3d ed. Philadelphia and London, 1924.
Gille, Bertrand. *Les origines de la grande industrie métallurgique en France*. Paris, 1947.
Godechot, Jacques. "L'aérostation militaire sous le Directoire," in *Annales historiques de la Révolution française*, VIII (1931), 213-28.
Goldstrom, John. *A Narrative History of Aviation*. New York, 1930.
La grande encyclopédie, inventaire raisonné des sciences, des lettres et des arts. Ed. by André Berthelot and others. 31 vols. Paris, 1886-1902.
Histoire de l'Académie royale des Sciences, années 1699-1790. 93 vols. Paris, 1702-1797. (More commonly known as *Mémoires de l'Académie des Sciences*.)
Hodgins, Eric, and Magoun, F. Alexander. *Behemoth: The Story of Power*. Garden City, N. Y., 1932.
Hunter, Dard. *Papermaking: The History and Technique of an Ancient Craft*. New York, 1943.
Inventaire-sommaire des archives départementales antérieures à 1790. Bouches-du-Rhône. Ed. by Louis Blancard. Paris, 1865.
——————. *Côte d'Or*. Ed. by Claude Rassignol and Joseph Garnier. Paris, 1864.
——————. *Gironde*. Ed. by J. B. Gras. Paris, 1864.
——————. *Ille-et-Vilaine*. Ed. by Edouard Quesnet and Paul Parfouru. Rennes, 1892.
Johns, W. E. *Some Milestones in Aviation*. London, 1935.
Kaempffert, Waldemar. *A Popular History of American Invention*. 2 vols. New York and London, 1924.

Kennelly, Arthur E. *Vestiges of Pre-Metric Weights and Measures Persisting in Metric-System Europe, 1926-1927*. New York, 1928.

Kiréevsky. *Histoire des législateurs chimistes: Lavoisier, Berthollet, Humphry Davy*. Frankfurt a. M., 1845.

Lacroix, Paul. *XVIIIme siècle: lettres, sciences et arts, France 1700-1789*. Paris, 1878.

Legras, P. Théodore. *Notice historique sur les deux hôpitaux et l'asile des aliénés de Rouen*. Rouen, 1827.

Levasseur, E. *Histoire des classes ouvrières et de l'industrie en France de 1789 à 1870*. 2d ed., 2 vols. Paris, 1903-1904.

Lodian, Walter. "A Century of the Telegraph in France," in *Popular Science Monthly*, XLIV (1893-1894), 791-801.

McCloy, Shelby T. *Government Assistance in Eighteenth-Century France*. Durham, N. C., 1946.

Machines et inventions approuvées par l'Académie royale des Sciences, depuis son établissement jusqu'à présent: avec leur description. Ed. by Gallon. 7 vols. Paris, 1735-1777.

Mémoires de l'Institut national des Sciences et Arts. Sciences mathématiques et physiques, pour les ans IV-XIII [1796-1805]. 6 vols. Paris, 1798-1806.

Mercier, L. S. *Tableau de Paris*. New ed., 12 vols. Amsterdam, 1782-1788.

Monfalcon, J. B. *Histoire monumentale de la ville de Lyon*. 9 vols. Paris and Lyon, 1866-1869.

Montross, Lynn. *War through the Ages*. Rev. ed. New York and London, 1946.

Nouvelle biographie générale depuis les temps plus reculés jusqu'à nos jours, avec les renseignements bibliographiques et l'indication des sources à consulter. Ed. by J. C. F. Hoefer. 46 vols. Paris, 1853-1866.

Ocagne, Maurice d'. "Un inventeur oublié: N.-J. Conté," in *Revue des deux mondes*, 8th ser., XXII (1934), 912-24.

Park, Joseph H., and Glouberman, Esther. "The Importance of Chemical Developments in the Textile Industry During the Industrial Revolution," in *Journal of Chemical Education*, IX (1932), 1142-70.

Partington, J. R. *A Short History of Chemistry*. London, 1939.

Power, D'Arcy, and Thompson, C. J. S. *Chronologia Medica: A Handlist of Periods and Events in the History of Medicine*. New York, 1923.

Procès-verbeaux du comité d'instruction publique de la Convention nationale. Ed. by James Guillaume. 6 vols. Paris, 1891-1907.

Rambaud, A. "Les sciences pendant la Révolution et l'Empire," in *Révolution française*, XIII (1887), 107-45.

Réimpression de l'ancien Moniteur [universel], seule histoire authentique et inaltérée de la Révolution française depuis la réunion des Etats-généraux jusqu'au Consulat (mai 1789–novembre 1799), avec des notes explicatives. Ed. by A. Ray. 32 vols. Paris, 1858-1863.

Rémond, André. *John Holker, manufacturier et grand fonctionnaire en France au XVIIIe siècle, 1719-1786.* Paris, 1946.

Richard, C. "Les savants et le salpêtre en Normandie sous la Terreur," in *Révolution française,* LXXVI (1923), 231-46.

Schmidt, Charles. "Les débuts de l'industrie cotonnière en France, 1706-1806," in *Revue d'histoire économique et sociale,* VII (1914-1919), 26-55.

Scoville, Warren C. "State Policy and the French Glass Industry, 1640-1789," in *Quarterly Journal of Economics,* LVI (1941-1942), 430-55.

Spaulding, Oliver Lyman, Jr.; Nickerson, Hoffman; and Wright, John Womack. *Warfare: A Study of Military Methods from the Earliest Times.* New York, 1925.

Stephens, Frederic George. *A History of Gibraltar and Its Sieges.* 2d ed. London, 1873.

Taylor, James A. *History of Dentistry: A Practical Treatise for the Use of Dental Students and Practitioners.* Philadelphia and New York, 1922.

Thompson, Charles J. S. *The History and Evolution of Surgical Instruments.* New York, 1942.

Timbs, John. *Wonderful Inventions: From the Mariner's Compass to the Electric Telegraph Cable.* London, 1867.

Torlais, Jean. *Un esprit encyclopédique en dehors de "l'Encyclopédie": Réaumur, d'après des documents inédits.* Paris, 1936.

Tuetey, Alexandre (ed.). *L'assistance publique à Paris pendant la Révolution.* 4 vols. Paris, 1895-1897.

────── (ed.). *Répertoire générale des sources manuscrites de l'histoire de Paris pendant la Révolution française.* 11 vols. Paris, 1890-1914.

Turgot, A. R. J. *Oeuvres de Turgot et documents le concernant, avec biographie et notes.* Ed. by Gustave Schelle. 5 vols. Paris, 1913-1923.

Usher, Abbott Payson. *A History of Mechanical Inventions.* New York, 1929.

Vaucanson, Jacques de. *Le mécanisme du flûteur automate, presenté à messieurs de l'Académie royale des Sciences . . .* Paris, 1738.

Vautier, Gabriel. "Les bullets incendiaires en 1793," in *Revue historique de la Révolution française,* IX (1918), 509-10.

Vierendeel, Arthur. *Esquisse d'une histoire de la technique.* 2 vols. Brussels and Paris, 1921.
Voltaire, François M. A. de. *Oeuvres complètes de Voltaire.* Ed. by Louis Moland. New ed., 52 vols. Paris, 1877-1885.
Witham, G. S., Sr. *Modern Pulp and Paper Making: A Practical Treatise.* 2d ed. New York, 1942.
Wolf, A. *A History of Science, Technology, and Philosophy in the Eighteenth Century.* New York, 1939.
Young, Arthur. *Travels in France by Arthur Young during the Years 1787, 1788, 1789.* Ed. by M. B. Betham-Edwards. 2d ed. London, 1889.

Index

Abeille, 162
Abrogast, 44
Académie royale des Sciences, character and function, 2-3
Achard, 81
adding machine, 123-24
Addison, Joseph, 41
age of inventors, 187 with n.
Albert, Antoine, 87
Alcock, 179
Alembert, D', 3, 195
Allonville, Jacques Eugène, Chevalier de Louville, portable transit, 127
ambulances, for soldiers, 166-67, 168
Amontons, Guillaume, 42 n.
André, l'aîné, 91 with n.
Anel, Dominique, 158
Angiviller, Comte d', 107
appendicitis, 157
Appert, Nicolas, on canning, 9, 81
Arcon, D', see Le Michaud
Argand, Aimé (also written Ami), his lantern, 52-56, 172 n.; collaborated on hydraulic device, 116-17
Arkwright, Richard, 91
Arlandes, Marquis d', 16
artificial limbs, 162-63
Audemar, 95
automata, 103-109
Auxiron, C. F. J., Chevalier d', work with steamboat, 31-33

Bachelier, Jean Jacques, painting in encaustic, 77; other activities, 78
Bacqueville, Marquis de, flight and fall, 26
Bailly, 51
balloon, 11-27; two types, 14; dirigibles, 23-25; military, 20-23
Ballooning Corps, 22, 23, 78
bandages, 162
Banks, Sir Joseph, 27
Baseilhac, Jean, known as Frère Côme, lithotomist, 155-56 with n.
Bayen, 147
Beaudeau, Abbé, 181
Beaudelocque, Jean Louis, 161

Beaumarchais, 126, 187 n.
bed, hospital, 166
Belidor, Bernard Forest de, "globes of compression," 143 with n.
Bellegarde, 144
Belloc, Jean Louis, 162
bellows, 122
Bernoulli, Daniel, 3, 30, 31, 118
Berthelot, Claude François, hand mill, 112; gun carriage, 140; renunciation of monetary returns, 189
Berthollet, Claude Louis, on scientific commission, 21; on bleaching with chlorine, 71-73; work with explosives, 73, 74, 147; on dyes, 82; on chemical nomenclature, 88; instructor to Widmer, 100; on Bureau of Consultation, 174; renunciation of monetary returns, 189
Berthoud, Ferdinand, 126, 175
Bessel, 127
Betancourt, 41
Biard, loom inventor, 97
Biberon, 112
Bichat, François Xavier, laid foundations of histology, 149
Biheron, Mlle, wax figures, 163
Blanchard, François, Channel flight, 17; parachuting of animals, 19; attempt at a dirigible balloon, 23-24; flying machine, 26-27
bleaching, 71-72
Bonnemain, heating of homes with hot water, 121; experimentation with incubator, 122, 123; devastating weapon, 146
Bonvallet, 98
Bordeu, Théophile de, on glands, 149-50
Bouceret, 95, 96
Bouguer, Pierre, 127
Bouillet, father and son, 145
Bourcet, Pierre de, 137
Bourgeois de Châteaublanc, Dominique François, oil lantern, 51, 106; claim to the invention of Vaucanson's duck, 106; lighthouse, 106-107

206 INDEX

Braille, 9
Bralle, François Joseph, 123
Bramah, Joseph, 124
Bréguet, Louis and Jacques, 9
brevets d'invention, in general, 7, 171, 172 n.; refused on steamboat, 31-32, 34-35; on distillation, 52, 59; refused on Argand lamp, 55; on Carcel lamp, 56; on Lebon's gas lamp, 59; on Robert's process for paper, 68, 69, 70; to Torré, 143
bridges, iron, 119; pontoon, 7, 140
Bridgewater, Duke of, 36
Brizout de Barneville, François Nicolas, invented spinning machine, 92
Brizout de Barneville, Nicolas Denis François (son of former), vicissitudes, 92-94; aided by government, 171
Brückmann, Franz Ernst, 66
Buffon, 108
Bureau of Consultation of the Arts and Trades, 173-74, 193

Cadet de Vaux, 53
Café Mécanique, 109-10
Calonne, 34
Camus, François Joseph, figurines, 106; crane, 113; pontoon bridge, 140; renunciation of economic returns, 189
canals, 35 with n., 119 n.
canning, 4, 81-82
cannon, 142
Carcel, Guillaume, lamp, 56-58
carding machines, 90-91
Carnot, Lazare, 21, 45, 79, 136
Carrochez, 128
Castel, Louis Bertrand, inventor of color-sound harpsichord, 131
cataracts, 158-59, 196
Catherine the Great, 3, 8, 107
Caylus, Marquis de, 77
Chabot, Joseph Xavier, 86
Chamoy, 113
Chappe, Claude, telegraph, 42-49
Chappe's brothers, 42, 47
Chaptal, 74, 82 n., 83 n., 85 n.,
Charles, J. A. C., use of hydrogen for balloons, 12-13, 15, 27, 189
Chaumette, De la, 142
Chauvet, 85
Chiarugi, Vincenzo, 151
chimney drafts, 121-22
China, claim to the balloon, 1 n.; manufacture of paper, 62; porcelain, 76; lacquer, 77; gilding, 78

Chipart, 129
chlorates, 74
Choiseul, Duc de, 37, 38, 137
chronometers, 125
Claude, Georges and André, 9
clavecin oculaire, 131-32
Cleland, 158
Clouet, Louis, on steel, 76
Côme, Frère, *see* Baseilhac
Condé, town captured, 45
Condorcet, 195
conscription, military and labor, 138-39
Conservatoire des Arts-et-Métiers, directed by Molard (1799), 38; created, 174-76
Conté, Nicolas Jacques, balloonist, 21, 23; pencilmaker, 78-79; biographical sketch, 78-80; economic indifference, 189
Coste, of Dauphinais, 144
Coste, Rev. J. P., of Charleston, on flame thrower, 144-45
Coudray, Madame Angélique Marguerite Leboursier de, on midwifery, 163-64, 165
Coulomb, Charles Augustin, inventor of torsion balance, 118-19 with n.
Coulon de Thévenot, fountain pen, 130-31
Courtivron, De, on inoculation of cattle, 152-53
cranes, 114-15
Crompton, 101, 182
Cugnot, Nicolas Joseph, steam truck, 7, 37-40; neglect by government, 39-40, 190
Cuisinié, harpsichord, 131
Curies, the, 9, 194

Daguerre, 9, 194
Daquin, Joseph, 151
Darcet, 80
Daunou, 44
David, Jean Pierre, 158
Daviel, Jacques, oculist, 158-59
Decène, Abbé, 95
Decrétot, manufacture at Louviers, 102 with n., 192
Delacourcière, 25
Delaunay, Léon, 47
Delmas, Louis, 177 n.
Delorme, 86
dentistry, 160-61
Desault, Pierre Joseph, surgeon, 149, 157, 161

INDEX

Desblancs, 35-36
Descroizilles, inventor of large-scale bleaching, 71, 72
Desforges, canon of Etampes, 26
Desmarest, Nicolas, authority on papermaking, 62 n., 63
Desquimare, 25
Desvallons, 114
Dickens, J., 30
Diderot, 131, 163, 195
Didot, Firmin, improved stereotyping, 129-30
Didot Saint-Léger, 68, 69-70
Didot publishing house, 68, 129-30
Diesbach, 83
Doulcet, Denis Claude, 153 with n.
dredges, 114
Droz, Henri Louis Jacquet, 108
Droz, Pierre Jacquet, 108
drugs, 153-54
Dubois, 114-15
Dubois-Crancé, 138
Dubois de Chémant, Nicolas, 161
Ducrest, 33
Dudit de Mezières, 115
Duhamel de Monceau, 121, 176
Dutrone, 80
dyes, 82-88, 98-100

ear trumpets, 95 n., 165
eau de Javelle, 72
education of inventors, 186-87 with n.
embargo on export of machines, by British, 182-83
encaustic painting, 77
Endelcrantz, 48
engineering, military and naval, 140-41
engraving tools, 129
Entreprenant, 22
Euler, Leonhard, 2, 30
Evans, Oliver, 5
explosives, 73-74, 145-47, 176
Eymar, merchant of Nîmes, 85, 86

Fabre, 144
factories in France before 1789, 191-92
Fauchard, Pierre, 160 with n.
Faujas de Saint-Fond, 12, 54
Favre, 50
Ferry, André, 115
Figuier, Pierre, 81
firearms, 142-46
first-aid service, on battlefield, 167, 196
Fitch, John, 5, 35
Fleurus, 22

floating batteries, 141
Folard, Chevalier, 136, 137
Follenay, Chevalier de, on steamboat, 31, 33, 34, 35
Foucault, Léon, 9
fountain pen, 130
Fourcroy, chemist, on commission of scientists, 21, 88; praise of the military balloon, 22; work with gunpowder, 74, 147; on committee for the pencil, 79 n.; popularized the new chemical nomenclature, 89; lecturer, 182 with n.
Fourdrinier, Henry and Seely, 70
Fournier des Granges, 90
Foxlow, 101, 178-79
Franklin, Benjamin, on electricity, 5; on use of French, 8; on balloons, 16, 27
Frederick the Great, 3, 8, 105
Fulton, Robert, 35-36

Gâche, De la, 112
Gamble, John, 70
Garnerin, A. J., exhibitions with parachute, 19
Garnet, 101, 182
Garnier, 146
Garrin, 119
gas lighting, 4
Gass, 146
Gautier, Abbé, 30
Ged, William and James, 130
Genevois, Swiss clergyman, proposed steamboat, 31, 33
Genoux, Claude, 67
Georget, 177
Germondy, Antoine, 90
Gibraltar, attack on, Sept. 13, 1782, 141
Gillot, Firman and Charles, 9
Girard, Philippe de, 129
glass, manufacture by British, 4; flint, 128; painted, 77, 179; French factory for, 192
Gondouin-Deshais, 116
Goudar, François, dye inventor, 85
Gouffé, Abbé, 114
Grandville, 91 with n.
"Greek fire," 143 with n., 144-45
Grenet, 100
Gribeauval, Jean Baptiste Vaquette de, high military consultant, 37, 136-37, 140
Guettard, Jean Etienne, 66
Guibert, Comte de, military tactician, 136-37

INDEX

Gullet, Léon, 9
Gurney, 40
Guyot, 158
Guyton de Morveau, on military value of balloons, 21; on chemical nomenclature, 88

Hahn, Matthew, 124
Hales, Stephen, on distillation of sea water, 80; on ventilator for ships, 120
Hargreaves, James, 91 with n., 94
Harrison, John, his chronometer, 125
heating of homes, by hot water, 121
heliometer, 127
Hellot, chemist, 83, 86
Hericé, 114
Herman, D', 140
Hildebrand, Frédéric, 95 with n.
Hillerin de Boistissandeau, 117-18, 123
Holker, John, English textile expert in France, 178
horology, 4, 124-26
Howard, 74-75
Huette, Louis, optician and discoverer of secret of flint glass, 128, 129
Hulls, Jonathan, 30, 36
Hunter, Alexander, 151
Hunter, Dard, 62 n.
Huntsman, 76
Huxley, 2
hydraulic devices, 115-17

incubation, of eggs, 122-23
Industrial Revolution, 7, 182-84
inoculation, 152-53

Jacquard, silk loom, 9, 98, 188
Jannin, creator of artificial pearls, 78
Jaubert, 97
Javelle, eau de, 71
Jeaurat, Edme Sebastien, 128
Jeffries, Dr., Channel flight, 18
Jenner, Edward, 151-52
Joubert, De, 52, 53, 54
Jouffroy d'Abbans, Marquis de, invention of the steamboat, 29, 32-34; failure to obtain *brevet d'invention,* 34-35; neglect by government, 190
Jourdan, J. B., 22
journals encouraging invention, 182
Jurine, 97, 98

Kay, John, 178
Kempelen, Wolfgang von, 108
Kircher, 131

lacquers, 7, 77
La Faye, De, 166-67
Laffecteur, 154
Lakanal, 44
Lalande, Joseph Jerome de, 21, 127-28
Lamanon, Paul, claims at ballooning, 25
Lange, piracy of Argand's lamp, 54-55
La Noue, François de, 162
Larrey, Jean Dominique, originator of first-aid service to soldiers, 166-67
Lasalle, Philippe de, loom, 96-97; hospital bed, 166
Laurent, Pierre Joseph, 162-63
Lavoisier, Antoine Laurent, lamp, 51; association with Argand, 52; participation in revision of chemical nomenclature, 88 with n.; on explosives, 147
Leblanc, Nicolas, creation of artificial soda, 75 with n.
Le Bon (probably to be identified with Philippe Lebon), 146-47
Lebon, Philippe, use of gas for lighting, 58-61
Le Cat, Claude Nicolas, lithotomist, 155, 156
Leckie, 111
Leclerc, 101
Le Conte, 132-33
Ledran, Henri François, 161
Le Gendre, Sandos, 124
Le Maire, 128
Le Masson, Gabriel, 115
Lemercier, 9
Lémery, Nicolas, 74, 147
Le Michaud, Jean Claude Eléonore, Chevalier d'Arçon, projector of attack on Gibraltar, 141
Lenoir, lieutenant general of police in Paris, 53, 54, 107
Lenormand, Sébastien, invented parachute, 19
Lepaute, Jean André, horologer, 126
Lepine, 123
Lepoule, 114
Leroy, Jean Baptiste, physicist and member of the Royal Society of Sciences, 51, 116, 125 n., 174
Leroy, Julien (father of Jean Baptiste and Pierre), noted horologer, 125
Leroy, Pierre, noted horologer, 125 with n., 126
Lerpold, Jacob, 124
Lesage, George Louis, 41, 53, 54
L'Espinasse, General, 145
Leturc, lace machine, 97

INDEX

levée en masse, 139
Levret, André, 161
Lhomond, technician, experiment at electrical telegraphy, 41; carding machines, 90, 91, 94; anticipation of the "spinning jenny," 94 with n.; fire grates, 121-22
Lhomond (son of former), 21
Liddell Hart, B. H., 23
Lind, James, 81
Linnaeus, 5
lithotomy, 153-56
locale, native, of inventors, 187-88 with n.
locks, safety, 124
looms, 96-98
Lot, 109
Louis XIV, 51, 74
Louis XV, 74, 78, 123, 143, 145
Louis XVI, 77
Louis, Marc Antoine, lithotomist, 155-56, 168
Louville, see Allonville
Lucas, plastic surgeon, 158
Luxeuil, 131
Lyttleton, 118

MacCloud, John, 179
Machine de Marly, 115 with n., 116
Mackay, 142
Macquer, Pierre Joseph, on dyes, 82-84; opportunity for research, 190
Mahon, Viscount (*later* Earl of Stanhope), 124
Maillard, figurines, 106
Maître-Jean, Antoine, 159
Malassagny, De, 111
Mallois, the brothers, 179
Malouet, 115
Mandre, Abbé de, 114-15
Maritz, Jean, cannon driller, 142-43
Marius, 132
Marre, 25
Marshall, 76
Martin, varnish, 77
Martin, Louis, 90-91
Massey, 101
Masson, Citoyenne, 67
mastoiditis, 156, 196
Mazéas, Abbé, 87 with n.
Meiffren, 111
Meikle, Andrew, 111
Menzies, Michael, 111
Mercier, L. S., 39
Merlin, François, 87
Mesnard, Jacques, 161

Mestivier, 157
metric system, 133-35, 196
Meudon, National Ballooning School, 23
Meusnier, J. B. M., balloon flight, 14; proposal of dirigible balloons, 24; lamp, 53 n.
Meynier, 117-18
Mical, Abbé, his figurines, 107
micrometer, 119, 127
microscopes, 126-28
midwifery, 164-65
Miller, Patrick, 35
mills, for grain, 112-13; for tobacco, 114
Milne, James and Thomas, 94, 101, 178, 179
Milot, 145
Milly, Count de, 54
Mittié, 154
Montalembert, Marquis de, military engineer, 140
Montgolfier, Etienne and Joseph, experimentation with balloons, 11-12; exhibition at Versailles, 14; cessation of contributions to balloon, 15; honors and popular fame, 16; witness of parachute, 19; collaboration with Argand on hydraulic device, 116-17; awarded medal for this device, 117; aided by government, 171
Morainville, 114
mordants, 65, 83-84 with nn.
Morgan, 101
Morrison, Charles, 41
motives, of inventors, 189
motor, for barges, etc., 114-15
"mule-jenny," 101
Müller, J. H., 124
Murdock, William, model steam cart, 40; use of coal gas for lighting, 60-61

Nanin, 147
Napoleon, letter from the Widow Montgolfier, 16; transportation of balloons to Egypt, 23; interest in the steam truck, 38; pensioning of Cugnot, 40; lukewarmness toward Chappe's telegraph, 46; request that Carcel exhibit lamp, 57; coronation day, 60; association with Berthollet, 73; praise of Conté, 80; award to Biard, 97; on application of the metric system, 135; adoption of Guibert's tactical feature, 137; admiration of Larrey, 167; use of flying ambulances in Italy, 168
Nef, John U., 2, 6

210 INDEX

Newcomen, Thomas, 28
Niepce, 9, 194
Nollet, Abbé, 122, 175
nomenclature, chemical, devised (1787), 88-89, 196
number of inventions, per individual, 187 with n.; per year, 192-93

Oberkampf, Christophe Philippe, factory at Jouy, 87, 99-100, 101, 192
odometers, 117-18
ophthalmology, 158-60
Orelly, 112
Orleans, Duc d', 73, 75, 107
Ormesson, D', 53
Oudet, 162
Outhier, Réginald, celestial sphere, 108

Pallouis, the Widow, producer of scarlet dye, 86-87
Panckoucke, Charles, 195
Papin, Denis, inventor of the steam engine, 28; life, 29; boat on the Fulda, 29-30
parachute, 19-20
parasol, 132
Passement, Claude Simon, 109, 128
Pasteur, 9, 194
patent law, of January 7, 1791, 171-72, 173; of March 29, 1791, 174-75
patents, in other countries than France, 173
Paulet, Jean, 97
Paulet, Paul, 95, 96
penalties for exporting machines, 183
pencils, 78-79
Percy, Pierre François, on first aid to wounded, 166-67, 168
Pereire, Jacob, 123-24, 187 n.
Périer, 33, 35, 116, 132, 163
Perkins, J. B., 75
Petit, François et Etienne du, 159
Petit, Jean Louis, surgeon, 156-57; screw tourniquet, 161
Petit, Marc Antoine, 149
Phillips, 56
philosophes, attitude toward invention, 194-96
Picard, 112
Pickford, Philemon, English textile expert in France, 101, 179
Pilâtre de Rozier, Jean François, flight in 1783, 15; pension, 16; fatal Channel flight, 17-18; director of "museum," 182

Pinel, Philippe, psychiatrist, 150-51
Pla, Simon, 90
Planta, 37
Poissonnier, 80
Pommier, 120
porcelain, 7, 76, 77
Porta, Jean Baptiste, incubator, 122, 123
Potter, Christopher, 179
Pourcheff, 117
Poux-Landry, Ambroise, 114
printing of cloth and paper, 7, 98-100, 177
printing press, 129
prizes, 51, 73 n., 147, 176, 181
Progin, Xavier, 9
Proust, 17
psychiatry, 150-51
pumps, 115-16
Puzos, Nicolas, obstetrician, 157 with n.

Quet, Du, 111, 162
Quinquet, pirated lamp, 54-55

Rabaut-Pommier, Jacques Antoine, 151-52
Rateau, Auguste, 9
Ravaton, 167
Réaumur, René Antoine Ferchault de, on papermaking, 65-66, 67; on steel, 75-76 with n.; on porcelain, 76; thermometer, 77 n.; report on artificial pearls, 78; improvements to the incubator, 122-23; pension, 190
Reisser, 41
Renard, 146
Ressin, Père, 113
Reveillon, 177 with n., 192
Rey, Jean, 9
Reynard, 116
Richer, micrometer, 119
Richet, Professor, 9
riots, destroying machines, 183-84
Rival, 95, 96
Rivarol, 107
Rivaz, 129
Rivey, of Rivet, 97
Rivey, Claude, of Paris, 98 with n.
Robert, François Nicolas, large-scale manufacture of paper, 7, 68-70
Robert Frères, 13, 14, 16
Robillard, printing machine, 99
Rocamus, 97
Rochon, Abbé Alexis Marie, 129
Roland, 38, 99
Rouillé de Meslay, 181

INDEX 211

Rozier, Abbé, 52
Rumsey, James, 5, 35

Saint-Sauveur, Jacques Grasset de, 166, 168, 189
Salva, Francisco, 41
Sarrazin, 90
Saulem, Rennequin, 115 n.
Savery, 28, 29
Saxe, Maréchal de, treadmill for boats, 30; on tactics, 137; carbine and cannon, 142
Schaffer, Christian, 67
Scheele, Karl Wilhelm, chemical contributions, 5; on bleaching with chlorine, 67 with n., 71
Scott, Baron de, 24
Seba, Albert, 66
Sebastian, Père, see Truchet
Seguin, Armand, 82 with n.
Selle de Beauchamp, Baron de, 22
shorthand, 130
Shrapnel, Major Henry, 143
Smellie, William, on midwifery, 164-65
Smith, Preserved, 2
Société Apollonienne, 2
Society for the Encouragement of Arts, Manufactures, and Commerce, 77 n., 86, 191
societies for the encouragement of invention in France, 178, 181-82
soda, 75
Soumille, Abbé, 129
spinning machines, 91-96, 101
Spooner, Eliakim, 5
steamboat, 5, 28-36
steam truck, 37-40
steel, 76, 176
Stephanopoli, Dimo, on black dye, 87
Stephenson, George, 40
Stevin, Simon, author of *The Decimal*, 133
Strada, Flamianus, 41
subsidies, government to inventors, 171, 176-78
Suchet, Etienne, 95
sugar, 80-81
Sully, Henry, horologer, 125-26
surgery, 155-69
surgical instruments, 161-62
surveyor's level, 129
Sutton, Thomas, 121
Symington, William, his steamboat, 36

Tabarin, 95-96

tachygraphy, 130
tactics, military, 136-37, 141
tanning, 82 with n.
telescopes, 126-28
Tellié, 90, 91
Tennant, Charles, 72
Teral, 122
textiles, English superiority in, 4, 7, 101-102, 178-82; French, 90-102; carding, 90-91; spinning, 91-96; weaving, 96-98; printed goods, 98-100; the factory system employed in, 102
textile experts and workers, imported, from England, 178-79; from Europe, 179-80; from the Orient, 180
threshing machine, 111-12
Torin (or Thorin), 78 with n.
Torré, inventor of incendiary bullet, 143
torsion balance, by Coulomb, 118-19
tourniquet, the screw, 161
Tremel, Jean, 114
Trevithick, Richard, 40
Trouville, 116
Truchet, Abbé (called Père Sebastian), automatic figurines, 105-106; made artificial hand, 162; invented ear trumpet, 165-66
Turgot, encouragement of invention, 73 with n., 96, 97, 129, 176-77 with nn.; suggested a metric system, 133

vaccination, 151-52
Vandermonde, 174
varnishes, 77, 78
Vaucanson, Jacques de, on the spinning machine of Villard, 95; own spinning machine, 96; member of committee examining loom, 97; inventor of silk loom, 97-98; automatic figures, 103-105, 106; aided by government, 171; left collection of machines to king, 174; inspiration, 189
Vauquelin, 74, 147
ventilators, for ships, hospitals, homes, 120-21
Vera, Charles Vincent, 116 with n.
Verniquet, 108
Villard, 95, 96
Villons, 142, 143
Voltaire, 194 with n., 195

watchmaking, 125-26
Watt, Gregory (son of James), 61

Watt, James, noted Scotch inventor, his relations to Murdock, 40, 61; proposal of screw propellers for ships, 118

Watt steam engine, 28, 34 n.; imported into France, 101

weaving, 96-98

Whitehurst, John, 117, 190

Widmer, Samuel, nephew of Oberkampf, 99; inventor of cylinder for printing goods, 100

windmill, 115

Young, Arthur, 3, 41 n., 102 n.